CHEMISTRY RESEARCH AND APPLICATIONS

# PHTHALATES

# HEALTH EFFECTS, DETECTION AND EXPOSURE PREVENTION

# CHEMISTRY RESEARCH AND APPLICATIONS

CHEMISTRY RESEARCH AND APPLICATIONS

# PHTHALATES

## HEALTH EFFECTS, DETECTION AND EXPOSURE PREVENTION

MIGUEL OWENS
EDITOR

nova publishers
*New York*

## Library of Congress Cataloging-in-Publication Data

Names: Owens, Miguel, editor.
Title: Phthalates : health effects, detection, and exposure prevention / Miguel Owens, editor.
Description: Hauppauge, New York : Nova Science Publishers, Inc., 2015. | Series: Chemistry research and applications | Includes index.
Identifiers: LCCN 2015037922 | ISBN 9781634839082 (hardcover)
Subjects: LCSH: Phthalate esters--Environmental aspects. | Phthalate esters--Health aspects. | Phthalate esters--Toxicology. | Esters. | Plasticizers.
Classification: LCC TD196.P47 P484 2015 | DDC 615.9/5138--dc23 LC record available at http://lccn.loc.gov/2015037922

*Published by Nova Science Publishers, Inc. † New York*

# CONTENTS

# PREFACE

Phthalates are ester derivatives that have been used to alter the flexibility of plastics since the 1930s, and are ubiquitous in our daily life. The first chapter of this book discusses how perinatal exposure to phthalates may induce alterations in brain neural growth and differentiation, with neurocognitive and behavioral consequences. Chapter two investigates the advantages and/or disadvantages of some analytical methods used in the past few years for chromatographic detection. The fourth chapter discusses the widespread application of phthalates and the role of liquid chromatography-mass spectrometry technique used in the detection, determination and confirmation of the chemical. The final chapter examines the assessment of phthalate exposure in polymer materials with advanced numerical algorithms.

Chapter 1 – Phthalates are ester derivatives that have been used to alter the flexibility of plastics since the 1930s, and are ubiquitous in our daily life. They can be transferred to the human being through air, ingestion and direct contact. There is increasing evidence suggesting that phthalates may act as endocrine disruptors, especially early in life, but concerns about neurodevelopment is more recent. In this chapter, the authors aimed to report the main available evidence about possible associations of perinatal exposure to phthalates and child neurodevelopment, reviewing the literature from *in vitro* to human epidemiologic data. *In vitro* studies have shown a dose-dependent effect of phthalate metabolites on the proliferation, differentiation and death of immature neuronal cells. Animal studies, mostly in rats, show some reduction in dendritic and synaptic formation, cell death through apoptosis, and markers suggestive of oxidative stress and inflammation, also in the hippocampus. Alterations in visuo-spatial learning and memory and in motor behavior and abilities have been observed. These effects appeared to vary according to the type of phthalates, the dose of the exposure, and the gender. Moreover, cognitive decline and histological alterations in prenatally exposed, old-aged rats, have recently suggested a potential role of phthalates in the development of Alzheimer's disease. Extrapolations from animal data to humans remain of course matter of debate. In humans, observational studies suggest an association of prenatal phthalate exposure with several adverse outcomes: reduced general cognitive ability, externalizing symptoms with hyperactivity, aggressiveness, disruptive behavior and inattention, mainly in boys, and internalizing behavior and anxiety trait in girls. However, there are an important heterogeneity and several limitations in these studies, limiting the interpretation of their results. In summary, the literature, from bench to bedside, suggests that perinatal exposure to phthalates may induce alterations in brain neural growth and differentiation, with neurocognitive and behavioral consequences. Available data are to be taken with caution, as they are based mainly on animal and epidemiological studies. Nevertheless, the growing body of evidences pointing in the same direction should encourage public health makers to sustain further research, as well as clear and feasible regulation on this potential "neuro-developmental disruptor."

Chapter 2 – Phthalates, chemicals easily released into the environment, are found to be ubiquitous. Although some differences in toxicity are well established, current data are not sufficient to allow elucidation of the range of toxicological differences. One of the most important issues for these compounds regards the analytical determination: many papers in literature can be found using chromatographic methods. This paper would like to show and investigate the advantages and/or the disadvantages of some analytical methods used these last 5 years. More than 150 papers have been highlighted and a critical revision is performed in relationship to complex matrices such as food, beverage and vegetables.

Chapter 3 – Phthalates (diesters of 1,2-benzenedicarboxylic acid) are a group of anthropogenic chemicals that are broadly employed in many household and personal care products, toys, and medical supplies, as well as plasticizers in various domestic and industrial products to increase the flexibility, durability, and transparency of plastics such as polyvinyl chloride (PVC) and cellulose acetate. The widespread use and the application of these compounds have resulted in their ubiquitous presence in the environment, and because they are classified by many countries (including EU and USA) as carcinogenic, mutagenic and toxic to reproduction, human exposure to phthalates is currently receiving considerable attention in political and scientific circles. These compounds are known to be endocrine disruptors, man-made chemicals that disrupt the endocrine system by competing with endogenous hormones by binding to receptors or by altering the synthesis and metabolism of these hormones. Epidemiological studies on human exposure to phthalates from different countries have shown that increased phthalate exposure might have associations with potential health risks such as allergic diseases in children such as asthma and eczema, reproductive effects (male fertility), precocious puberty in pubertal girls, increased incidence of chronic disease, and a possible role in the development of cancer. Since 14 December 2005, the EU Directive 2005/84/EC has banned several phthalates for use in PVC and other plasticized materials in all toys and childcare articles. Moreover, several studies have concluded that diet is the major route of exposure for humans, and that environmental contamination is one of the sources of these chemicals in food at various levels. So, the development of analytical methodologies for the monitoring of these compounds in a great variety of matrices including environmental ones, food products and food packaging materials, as well as in biological fluids for epidemiological studies on human exposure is mandatory. Various analytical methods have been developed for measuring phthalates, mostly using gas chromatography coupled with mass spectrometry (GC-MS) and liquid chromatography coupled with mass spectrometry (LC-MS) or tandem mass spectrometry (LC-MS/MS). LC-MS(/MS) methods, which combine high selectivity and sensitivity, show an advantage over GC-MS methods because of simple sample preparation without the need for derivatization. Recently, high resolution mass spectrometry (HRMS) is acquiring also an important role in the analysis of phthalates by taking advantage of the higher resolving power and higher accurate mass measurements achieved, and today several LC-HRMS methods are reported for the quantitative determination of these compounds. In this chapter the role of liquid chromatography-mass spectrometry in the detection, determination and confirmation of phthalates in a variety of matrices (environmental, food and bio-analytical ones) will be reviewed. LC-MS(/MS) chromatographic conditions, ionization sources, and MS and HRMS analyzers frequently used, as well as strategies for the structural characterization and the qualitative and quantitative analysis of phthalates and their metabolites will be discussed by means of relevant applications. Coverage of all kind of applications is beyond the scope of the present contribution, so the authors will focus on the most relevant applications published in the last years.

Chapter 4 – Exposure of phthalates in polymer materials was assessed by employing migration modeling with advanced numerical algorithms. A modeling approach originally developed in Europe for assessing food contact materials for regulatory compliance was employed to assess the exposure risk of phthalates in public health. Migration estimation of phthalates was compared with the empirical data; the results of the migration modeling were within the allowable tolerance, thus proving its applicability to any contact medium, including human skin. In this respect, this migration estimation scheme can also be used for human exposure modeling to assess the exposure risk of phthalates. The new technique described above would enable effective assessment of exposure risks.

In: Phthalates
Editor: Miguel Owens

ISBN: 978-1-63483-908-2


*Chapter 1*

# PHTHALATES AND NEURODEVELOPMENTAL TOXICITY: A REVIEW OF THE EVIDENCE

***Myriam Bickle Graz, Jean-François Tolsa**[*]*
***and Céline Fischer Fumeaux***
Clinic of Neonatology, Department of Pediatrics, University Hospital of Lausanne, Lausanne CHUV, Switzerland

## ABSTRACT

Phthalates are ester derivatives that have been used to alter the flexibility of plastics since the 1930s, and are ubiquitous in our daily life. They can be transferred to the human being through air, ingestion and direct contact. There is increasing evidence suggesting that phthalates may act as endocrine disruptors, especially early in life, but concerns about neurodevelopment is more recent.

In this chapter, we aimed to report the main available evidence about possible associations of perinatal exposure to phthalates and child neurodevelopment, reviewing the literature from *in vitro* to human epidemiologic data.

*In vitro* studies have shown a dose-dependent effect of phthalate metabolites on the proliferation, differentiation and death of immature neuronal cells. Animal studies, mostly in rats, show some reduction in

[*] Corresponding Author: Pr. Jean-François TOLSA, Clinic of Neonatology, Department of Pediatrics, Av. Pierre Decker 2, University Hospital of Lausanne, CH-1011 Lausanne CHUV, Switzerland. E-mail: Jean-Francois.Tolsa@chuv.ch.

dendritic and synaptic formation, cell death through apoptosis, and markers suggestive of oxidative stress and inflammation, also in the hippocampus. Alterations in visuo-spatial learning and memory and in motor behavior and abilities have been observed. These effects appeared to vary according to the type of phthalates, the dose of the exposure, and the gender. Moreover, cognitive decline and histological alterations in prenatally exposed, old-aged rats, have recently suggested a potential role of phthalates in the development of Alzheimer's disease. Extrapolations from animal data to humans remain of course matter of debate.

In humans, observational studies suggest an association of prenatal phthalate exposure with several adverse outcomes: reduced general cognitive ability, externalizing symptoms with hyperactivity, aggressiveness, disruptive behavior and inattention, mainly in boys, and internalizing behavior and anxiety trait in girls. However, there are an important heterogeneity and several limitations in these studies, limiting the interpretation of their results.

In summary, the literature, from bench to bedside, suggests that perinatal exposure to phthalates may induce alterations in brain neural growth and differentiation, with neurocognitive and behavioral consequences. Available data are to be taken with caution, as they are based mainly on animal and epidemiological studies. Nevertheless, the growing body of evidences pointing in the same direction should encourage public health makers to sustain further research, as well as clear and feasible regulation on this potential "neuro-developmental disruptor."

## Abbreviations

| | |
|---|---|
| ADD | Attention deficit disorder |
| ADHD | Attention deficit and hyperactivity disorder |
| ASD | Autistic spectrum disorder |
| BASC | Behavior Assessment System for Children-Parent Rating Scales, parental questionnaire |
| BSID | Bayley Scales of Infant development, assessment tool of child development, and the second edition gives a mental developmental index (MDI) and a psychomotor developmental index (PDI). |
| BRIEF | Behavior Rating Inventory of Executive Function, a questionnaire of 86 questions encompassing 8 clinical scales of executive function. |

| | |
|---|---|
| CBCL | Child Behavior Checklist, parental questionnaire, preschool or childhood version. |
| CPT | Continuous Performance Test, computer based test of inattention and impulsivity. The number of errors (omission and commission) is recorded. |
| DBP | dibutyl phthalate |
| DEHP | Di(2-ethylhexyl)phthalate |
| DD | Developmental Disabilities |
| DMP | dimethyl phthalate |
| DOHaD | Developmental Origins of Health and Disease |
| FIQ | full-scale intelligence quotient |
| HMWP | high molecular weight phthalate |
| IQ | intelligence quotient |
| LD | learning disability |
| LMWP | low molecular weight phthalate |
| MEHP | mono-(2-ethylhexyl) phthalate |
| MRI | Magnetic Resonance Imaging |
| PIQ | Performance Intelligence Quotient |
| PND | postnatal day |
| PVC | polyvinyl chloride |
| REACH | Registration, Evaluation, Authorization and Restriction of Chemicals (REACH) |
| SDQ | Strengths and Difficulties Questionnaire, a parent/teacher completed questionnaire regarding the child's behavior. |
| WPPSI-III | Wechsler Assessment for Preschool and Primary School, third edition. |
| WISC-IV | Wechsler Intelligence Scales for Children, fourth edition. |

## INTRODUCTION

The neurodevelopment of children follows a relatively defined and straightforward path. Starting at conception, many impediments may arise, and many factors play an important role in a child's cognitive and behavioral abilities, as well as in general health. Known important factors are, for example, the parental socio-economic status, genetic anomalies, gestational age, or birth weight. The rising field of Developmental Origins of Health and Disease (DOHaD) explores the role of maternal and environmental factors during pregnancy and infancy in the long-term outcome of children. Research in this field is of great interest, as many of these environmental exposures could be modified in order to improve the long-term health and development. However, one of the main difficulties when studying the impact of one possible modifier is to take into consideration all of the other confounding factors that are often interrelated in the case of neurodevelopmental outcomes.

Of the different environmental factors supposed or known to have an impact on child neurodevelopment, phthalates have received an increasing attention in recent years. This chapter focuses on the potential neurobehavioral impact of perinatal exposure to phthalates, based on a review of the literature.

Phthalates have been used to increase the flexibility of plastics, particularly of polyvinyl chloride (PVC) polymers, since the 1930s, with an annual worldwide production now exceeding 4 tons/year. Based on their molecular weight, they can be classified into 2 groups, corresponding to different uses in the industry. The main phthalates and use thereof are summarized in Table 1.

**Table 1. Classification of main phthalates**

| Molecule | Abbreviation | Use |
|---|---|---|
| *High molecular weight phthalate (HMWP)* | | |
| Di-2-ethylhexylphthalate | DEHP | Plasticizer, polyvinyl chloride (PVC), vinyl flooring, packaging |
| Benzylbutyl phthalate | BzBP | |
| Di-n-octyl phthalate | DOP | |
| Di-isononyl phthalate | DiNP | |
| Di-isodecyl phthalate | DiDP | |
| *Low molecular weight phthalate (LMWP)* | | |
| Dibutyl phthalate | DBP | Personal care products, tablet coatings, aerosols |
| Dimethyl phthalate | DMP | |
| Diethyl phthalate | DEP | |

Phthalates are found in many common household items, including but not exclusively, flooring, wallpaper and ceilings, packaging, personal care products and pills, clothing and footwear. Phthalates are thus ubiquitous, and can be measured everywhere in the outdoor and indoor environment. Phthalates are transferred to humans by inhalation (of dust for example), by ingestion (dairy products, packaged food) and by direct contact (skin lotions, rain clothing). Indeed, phthalate metabolites have been measured in different body fluids, such as blood and urine, amniotic fluid [1], cord blood [2] and adipose tissue.

Concerns about the toxicity of phthalates have emerged since the 1990s. Some of them have been classified as possibly teratogenic, carcinogenic or toxic regarding the reproduction (groups IIB/III according to the the International Agency for Research on Cancer (IARC)/WHO classificationin 2000 [3]. The use of Di(2-ethylhexyl)phthalate (DEHP) in toys and childcare articles is prohibited in the European Union since 2006, and in food packaging since 2008. The European Parliament and the Council of the Registration, Evaluation, Authorization and Restriction of Chemicals (REACH) decided that as of 4/11/2015, phthalates considered toxic for reproduction may only be used after an authorization has been granted.

The presence of DEHP in numerous medical devices has also raised questions and concerns about possible noxious long-term effects in adults and especially in children. This medical exposure could be particularly critical in neonatal intensive care units, where sick newborns are both the most exposed to DEHP, and the most immature and vulnerable [4]. In 2007, the European Council issued a press release about a change in regulation, which stipulated that, "Devices which could possibly release phthalates to the body of the patient should be labeled accordingly." While this regulation was supposed to be implemented by 2011, medical devices containing phthalates can remain difficult to identify in practice [5]. In France, from the first of July 2015, all phthalate-containing devices are forbidden of use in maternity wards, as well as in neonatal and pediatric units.

In adults, phthalates have been inconsistently associated with obesity, diabetes, hypertension, and reproductive difficulties in men and women [6, 7].

In children, the number of studies about deleterious effects of phthalates has tremendously increased during the last 5 years.

These noxious effects are mainly reported to affect the endocrine and reproductive system, especially in males, with a feminizing effect (shorter ano-genital distance, and possibly testicular dysgenesis) [8-10]. Epidemiological studies in children have also described an association of prenatal and postnatal phthalates exposure with decreased gestational length and birth weight [11], the development of asthma [12], eczema and allergies [13], obesity [14], and higher blood pressure [15, 16].

However, most of the recent literature in children is about the associations of pre- and postnatal exposure to phthalates with various aspects of child neurodevelopment. Recently, two reviews were published on this topic [17, 18]. Here, we will first look at the available evidence coming from *in vitro* studies, then from animal studies and finally, from human epidemiological studies, for both low and high molecular weight phthalates. To perform this review, we performed a *pubmed* search with "children" and "phthalate" as key words, and completed with the reference lists from the published studies. We looked at references studying neurocognitive and neurobehavioral outcomes.

## *In Vitro* Studies of Phthalate and Mammal Brain Tissue

The first studies on phthalate toxicology were conducted on fish. Later, several authors studied the changes in mammal brain cell morphology and viability following exposure to phthalates (Table 2). The first *in vitro* study reporting neurotoxic effects of phthalates was published in 1974 by Kasuya, who exposed immature rat brains (neurons, fibroblasts and glia) to DEHP, dibutyl phthalate (DBP) and dimethyl phthalate (DMP), and showed alterations in the growth of dendrites and glia in the immature rat neurons [19]. Recently, Lee et al. also demonstrated inhibition of dendrite formation and cell death with DBP at high concentrations [20]. Chen et al. later showed that mono-(2-ethylhexyl) phthalate (MEHP) exposure was associated with a dose dependent arrest of differentiation in undifferentiated cells, and with cell death in differentiated neuronotypic PC12 cells [21]. Different authors have investigated the mechanisms of these effects. Aung et al. and Guida et al. demonstrated caspase-3-mediated apoptosis in DEHP-exposed cells of mice as well as human neuroblastoma and rat embryo cortical cells [22, 23]. However, these results were not confirmed by Ishido et al. who exposed rat embryo mesencephalic cells to 7 different phthalates [24]. They found that the 7 molecules inhibited cell migration, and that 5 molecules inhibited cell differentiation, but they did not find evidence of apoptosis. Finally, in 2014, Wu et al. used a co-culture of cortical neurons and astrocytes, and showed signs of dose-dependent oxidative stress and death in the neurons, and astrocyte proliferation and gliosis [25].

These *in vitro* effects were observed in rodent and human neurons, after exposure to high and low molecular weight phthalate (HMWP and LMWP). Moving forward from these *in vitro* studies, the way to enhance knowledge about the potential neurotoxicity of phthalates is through *in vivo* animal studies, which are yet ongoing.

**Table 2. Examples of *In vitro* studies of phthalate and mammal brain tissue**

| Authors (ref) | Year | Molecule(s) | Cell | Effect |
|---|---|---|---|---|
| Kasuya et al. [19] | 1974 | Di-n-butyl-phthalate Diethyl phthalate dimethylphthalate | P6-12 rat brains Neurons Fibroblasts glia | Dose dependent inhibition of growth: -dendrites -glia |
| Aung et al. [22] | 2014 | Di(2-ethylhexyl)phthalate | Mouse neuroblastoma neuro-2a | Dose-dependent caspase-mediated apoptosis |
| Chen et al. [21] | 2011 | Mono(2-ethylhexyl)phthalate | Neuronotypic PC12 cells, undifferentiated and differentiating | Dose dependent Cell cycle arrest, death Altered differentiation |
| Guida et al. [23] | 2014 | Di(2-ethylhexyl)phthalate | Human neuro-blastoma cells Rat embryo cortical neurons | Dose dependent reduction cell viability Apoptosis (caspase 3) |
| Ishido et al. [24] | 2014 | 7 different molecules | Rat embryo mesencephalic neurons | Inhibition of cell migration (7 mol) Inhibition of cell differentiation (5 mol.) - No apoptosis |
| Lee et al. [20] | 2015 | Dibutylphthalate | Rodent hippocampal neurons | Altered neurite formation Cell death at high concentration |
| Wu et al. [25] | 2014 | Di(2-ethylhexyl)phthalate | Mouse astrocyte-cortical neurons co-culture | Dose-dependent oxidative stress Dose-dependent neuronal death Astrocyte proliferation and gliosis |

## ANIMAL STUDIES OF PHTHALATES AND NEURODEVELOPMENT

Animal studies have mostly used rodent models, specifically rats, with a prenatal and a post natal exposition up to postnatal days (PND) 7-10, which corresponds to prenatal exposition in humans [26, 27].

**Table 3. Examples of animal studies of phthalates and mammal brain**

| Authors (ref) | Year | Route and Molecule | Exposure time | Effect(s) |
|---|---|---|---|---|
| Andrade et al. [28] | 2006 | Enteral DEHP | Rat dams: G6-L21 | Pups: Dose, gender and timing dependent altered brain aromatase activity |
| Dabholkar et al. [29] | 1988 | Enteral DEHP | Rat dams | Pups: peroxysomal proliferation in cortex |
| Ishido et al. [30] | 2004 | Intracisternal dicyclohexyl-phthalate | P5 male rat pups | Increased motor activity<br>Altered brain gene expression |
| Li X. et al. [31, 32] | 2013, 2014 | Enteral DBP | Rat dams: G6-P21 | Hippocampus apoptosis, ↓spatial memory<br>Down and up-regulation of signaling molecules |
| Li Y. et al. [33] | 2010 | Enteral DBP | Rat dams: G6-PND21 | PND 22 pups:<br>Improved learning at high dose DBP |
| Min et al. [34] | 2014 | Enteral BBP | Adult mice | High dose BBP:<br>↓Learning and memory<br>Brain oxidative damage<br>Down regulation of signaling proteins and serotonin |
| Peng et al. [35] | 2015 | Enteral DiNP | Adult mice | Dose-dependent:<br>↓spatial learning and memory<br>Oxidative stress, inflammation and apoptosis |
| Smith et al. [36, 37] | 2011 2014 | Intra-peritoneal DEHP | Rats PND 16-22 | Males:<br>↓connectivity and development of hippocampus<br>↓ neurochemical signaling |
| Sun et al. [38] | 2014 | Enteral DEHP | Rat dams G2 until weaning | At 21 months:<br>↓spatial learning and memory<br>Elevated phosphorylation of protein Tau |
| Tanaka et al. [39] | 2002 | Enteral DEHP | Mice dams and pups | Gender and dose-dependent delayed surface righting |

G: Gestational day; L: Lactation day; PND: Postnatal Day.

Most studies reporting on the association between phthalates and neurodevelopment have used rodents with a perinatal exposure to HMWP or LMWP. As shown in Table 3, phthalates have mostly been given repeatedly through the enteral route, although intracisternal and intraperitoneal injections where also tested. Study designs encompass behavioral paradigms to assess visuo-spatial learning and memory, as well as brain histology and immunohistochemistry in the exposed pups.

## Perinatal Exposure and Learning

Impaired learning and memory after perinatal exposure to LMWP has been found by Xiujuan Li et al. who gave 500 mg/kg/day intragastric DBP to rat dams, from gestation day 6 to PND 21 [40]. In their first study [40], they showed that, after DBP exposure, spatial learning was altered on PND22 pups. This was associated to apoptosis and neuronal cell loss in the hippocampus on PND 5 and 21, *but not in the mature animal* (day 60). In their second study [31], they performed immunohistochemical studies of the brain and western blots in the hippocampus. The authors observed significant differences between treated and control rats on PND 21 (up-regulation of aromatase and down regulation of other proteins in DBP treated animals), *but not at 60 days.* These studies suggest that DBP may induce neurotoxic effects in immature animals, markedly in the hippocampus, associated with a short-term impairment of learning and memory. Interestingly, these findings are contradictory to the ones reported by Li et al. who showed that after a high dose perinatal DBP exposure in rats, there was enhanced spatial memory and learning in PND22 and PND 62 pups [33]. This was accompanied by enhanced expression of the hippocampal brain-derived neurotrophic factor on PND 21.

Long term cognitive dysfunction (spatial learning and memory), and hyper-phosphorylation of protein Tau, a hallmark of Alzheimer's disease (without amyloid plaques) were reported by Sun et al. [38] in the hippocampus of rats aged 21 months that had been perinatally exposed to oral DEHP. They suggested that phthalates might play a role in the pathogenesis of Alzheimer's disease.

## Adult Exposure and Learning

Two authors who subjected adult mice to enteral HMWP also found this association with learning and memory. Min et al. exposed adult mice during 14 days to 1250 mg/kg/day of BBP [34], whereas Peng et al. exposed adult mice to increasing doses of DiNP [35]. They both observed decreased learning and memory, dose-dependent alterations in the hippocampus. Min et al. showed modified levels of hippocampus serotonin with signs of oxidative stress, whereas Peng et al. showed signs of oxidative stress (reactive oxygen species, gluthation, superoxide dismutase), inflammation (interleukin 1, tumor necrosis factor-α), and apoptosis (caspase 3) [35].

## Perinatal Exposure and Motor Activity

The quantity of motor activity may also be affected by phthalates, as Ishido et al. have demonstrated in male rats subjected to a single intracisternal injection of dicyclohexylphthalate [30]. Furthermore, Tanaka et al. observed a gender effect in the quality of motor activity after they had exposed pregnant mice and their pups to oral DEHP at different doses: females had worse scores at low doses of DEHP, whereas males were hindered by high doses of DEHP [39]. They later repeated the experiment, and mated the 9-week-old offspring with either a control group mouse (T/C), or another treated mouse (T/T). The males of these T/T litters moved significantly more [41].

Indeed, most of the animal studies have proposed the hippocampus as the site of the brain injury, with a histology that is similar to the *in vitro* results [36, 37].

The supposed mechanisms for these possibly long lasting alterations of neurodevelopment in animals exposed either perinatally, or as adults, to LMW and HMW phthalates could involve changes in aromatase activity. Andrade et al. studied brain aromatase activity in the hypothalamic and preoptic regions [28]. They subjected rat dams to oral DEHP in doses mimicking human exposure, from gestation day 6 to lactation day 21. Results showed timing-, sex-, and dose-related changes in aromatase activity, which could lead to altered brain differentiation and behavior.

Oxidative stress is another possible mechanism, and has been explored by several authors. In 1988, Dabholkar et al. showed an increase in the number of neurons containing peroxisomes in the PND 12 pups of DEHP fed rat dams, as well as an increase of the size of those peroxisomes [29]. More recently, Min et al. and Peng et al. found evidence for oxidative stress-related damage in the hippocampus [34, 35].

In summary, research using rodents shows neurotoxic effects of phthalates, which varies according to the time or the dose of the exposure, and the gender of the exposed animal, with a specific vulnerability of the hippocampus. Histology confirms the information obtained from *in vitro* studies: some reduction in dendritic and synaptic formation, cell death through apoptosis, and markers suggestive of oxidative stress and inflammation [36, 37]. In rodents, this cellular brain damage leads to impairments of spatial learning and memory in the first days of life. Interestingly, these changes were significant in the short-term, but seemed to disappear at around 60 days, where recovery might have taken place. However, the only long-term experiment showed behavioral and histological signs consistent with Alzheimer's disease in adult rats perinatally exposed to DEHP [38]. The previously mentioned recovery could thus be a transient remission.

While extrapolations from animal data to humans remain limited and matters of debate, we shall continue with the discussion of human epidemiological studies on phthalates and child neurodevelopment. Epidemiological studies linking phthalate exposure with child neurodevelopment.

## Epidemiological Studies Linking Phthalate Exposure with Child Neurodevelopment

Child neurodevelopment follows a fairly straightforward path. Developmental milestones are well known to parents and pediatricians alike, and give indications about the "normality" of an individual child's development. In order to measure a pediatric population's development, and to use these measures as outcomes in research, standardized, validated, reliable, and reproducible instruments are necessary.

As a child grows, the range of his abilities expands and the precision of the measures increases. We shall thus report on the studies examining neonates and babies, toddlers, and children, using different and appropriate direct tests or validated questionnaires.

**Table 4. Examples of human studies of phthalates and neurodevelopment**

| Authors (ref) | Year | Molecules | Assessment | Age at outcome | Number of patients | Main results |
|---|---|---|---|---|---|---|
| Engel et al. [42] | 2009 | MU: 10 metabolites (6 HMWP, 4LMWP) | Brazelton Neonatal Behavioral Assessment Scale (7 clusters) | Neonates < 5 days | 295 | Girls: HMWP:<br>↓ orientation and quality of alertness score.<br>Boys: LMWP: ↑ motor performance |
| Yolton et al. [43] | 2011 | MU: 6 DBP and DEHP at 16 and 26 weeks pregnancy | NICU Network Neurobehavioral Scale | 5 weeks | 350 | DBP at 26 weeks: ↓arousal and ↓ need for handling<br>DEHP: ↑ non-optimal reflexes in boys |
| Kim Y. et al. [44] | 2011 | MU: DEHP metabolites | BSID, 2nd edition | 6 months | 460 | Boys: ↓ MDI and PDI. |
| Polanska et al. [45] | 2014 | MU and CU at 24 months<br>5 LMWP and 6 HMWP | BSID, 3rd edition | 23-28 months | 165 | Several prenatal metabolites ↓motor score, not cognitive or language scores<br>No association for postnatal exposure |
| Tellez-Rojo et al. [46] | 2013 | MU: 9 metabolites (4DEHP) | BSID, 2nd edition | 24-30 and 36 months | 136 | Girls: DEHP:↓ MDI<br>Boys: LMWP:↑PDI |
| Whyatt et al. [47] | 2012 | MU: 4 DEHP metabolites and 3 non DEHP metabolites | BSID, 2nd edition<br>CBCL | 36 months (24-42) | 296 | Non DEHP: ↓PDI,<br>Girls:↓ MDI,<br>↑ internalizing behavioral problems<br>DEHP: no significant association |

| Authors (ref) | Year | Molecules | Assessment | Age at outcome | Number of patients | Main results |
|---|---|---|---|---|---|---|
| Gascon et al. [48] | 2015 | MU: 12 and 32 weeks<br>8 phthalate metabolites (5 HMWP, of which 4 DEHP, and 3 LMWP) | 1 year: BSID<br>4 years: McCarthy Scales of Children's abilities<br>California Social Competence Scale<br>DSM-IV based ADHD diagnosis<br>7 years: SDQ<br>Conner's Parent Rating Scales | 1, 4 and 7 years | 367 | No association with cognition at 1 or 4 years old<br>DEHP ↑ social competence, ↓ inattention, ↓ ADHD<br>LMWP (MEP)<br>↓ inattention<br>No gender effect |
| Engel et al. [49] | 2010 | MU: 10 metabolites (6 HMWP, 4LMWP) | BRIEF<br>BASC-Parent Rating Scales | 4-9 years | 188 | LMWP: ↓ executive functioning (BRIEF)<br>↑ attentional, aggressive, conduct problems (BASC) |
| Chopra et al. [50] | 2014 | CU with ADD, LD, ADD + LD, and controls<br>DEHP and DBP metabolites<br>MBzP, MEP, MCPP (5 HMWP, 4LMWP) | Parental questionnaire on diagnosis | 6-15 years | 1493 | DEHP: ↑ADD |
| Factor-Litvak et al. [51] | 2014 | MU, CU year 3 and 5: metabolites of 5 phthalates | WISC-IV | 7 years | 328 | CU phthalates: no association<br>MU phthalates: ↓IQ, processing speed, perceptual reasoning, working memory<br>-association stronger in girls |

**Table 4. (Continued)**

| Authors (ref) | Year | Molecules | Assessment | Age at outcome | Number of patients | Main results |
|---|---|---|---|---|---|---|
| Huang et al. [52] | 2015 | MU and CU: 7 metabolites including 3 DEHP metabolites | BSID<br>WPPSI-R<br>WISC-IV | 2-3, 5-6, 8-9, 11-12 years | 110 | MU phthalates: no association<br>CU phthalates: ↓IQ |
| Cho et al. [53] | 2010 | CU: 2 DEHP and 1 DBP metabolites | KEDI-WISC | 8-11 years | 621 | DEHP: ↓vocabulary |
| Kim B. et al. [54] | 2009 | Cross-sectional<br>CU:<br>2 DEHP metabolites, 1 DBP metabolite | -DSM-IV based -Teacher ADHD questionnaire<br>-CPT<br>-KEDI-WISC | 8-11 years | 261 | DEHP metabolites:<br>↑inattention symptoms<br>↑hyperactivity symptoms |
| Kobrosly et al. [55] | 2014 | MU (26 weeks)<br>6 metabolites (3DEHP, 2 DBP) | CBCL | 6-10 years | 153 | DEHP:<br>Boys:↑ somatic complaints<br>Girls: ↓ anxiety and depression<br>DBP: ↑attention, aggressive, conduct problems in boys |
| Lien et al. [56] | 2015 | MU (26 weeks) (3 DEHP, 4 DBP) | CBCL | 8-9 years | 122 | DEHP and DBP:<br>↑ externalizing, delinquent and aggressive scores |
| Park et al. [57] | 2015 | CU<br>DEHP metabolites (2)<br>DBP metabolite (1) | TAIC | 8-11 | 277 | DEHP metabolites:<br>↓ anxiety in girls only |
| Park et al. [58] | 2014 | CU non treated ADHD, DEHP and DBP metabolites | CPT | 6-15 years | 179 | DEHP and DBP metabolites:<br>↑ commission errors<br>Interaction with gene for dopamine receptor D4 |

| Authors (ref) | Year | Molecules | Assessment | Age at outcome | Number of patients | Main results |
|---|---|---|---|---|---|---|
| Park et al. [59] | 2014 | CU non treated ADHD, Control group<br>DEHP and DBP metabolites | Parent ADHD rating scale<br>Brain MRI T1 weighted | 6-15 years | 115 (MRI)<br>78/438 (case control) | DEHP: ↓cortical thickness right temporal<br>Case/control: ↑ phthalates in cases |
| Philippat et al. [60] | 2015 | House dust 2-5 years after birth in autistic spectrum (ASD), developmental delay (DD), and control (C) | ADI/ADOS<br>Mullen Scales of early Learning<br>Vineland Adaptive Behavior Scales | 2-5 years | 50(ASD)<br>27(DD)<br>68 (C) | DEHP and BBzP: ↑DD<br>In ASD and DD: DBP and DEP: ↑hyperactivity and impulsivity |
| Testa et al. [61] | 2012 | CU: 4 DEHP metabolites in ASD and controls | | 11 +/- 5 years | 48 (ASD)<br>45 (C) | ASD: ↑ ¾ DEHP metabolites |

MU: Maternal Urine; CU: child urine; Other abbreviations: *see above*.

Two studies report on neurodevelopmental outcome associated with phthalate exposure in neonates. Engel et al. measured 10 phthalate metabolites (4LMWP and 6 HMWP) in the urine of 295 third trimester pregnant women and examined the neonates before the fifth day of life with the standardized Brazelton's Neonatal Behavioral Assessment Scale, which gives results for 7 clusters of items (habituation, orientation, motor, range of state, regulation of state, autonomic stability, and number of abnormal primitive reflexes) [49]. They observed an inverse and significant correlation of HMWP metabolites concentration with the orientation and quality of alertness score in girls, and a slight association of increased LMWP levels with better motor performance in boys. In 2011, Yolton et al. did a study using two pregnancy time points (16 and 26 weeks) DEHP and DBP metabolites and the NICU Network Neurobehavioral Scale (NNNS) at 5 weeks of age [43]. This scale provides 13 sub-scores: habituation, attention, arousal, self-regulation, need for special handling, movement quality, excitability, lethargy, non-optimal reflexes, asymmetrical reflexes, hypertonicity, hypotonicity, and stress/abstinence (the last 2 were omitted, as they are seldom used). None of the measures at 16 weeks was associated with the NNNS. The DBP (LMWP) metabolites at 26 weeks were associated with better behavioral organization, less arousal and need for handling, whereas the DEHP (HMWP) metabolites were associated with less optimal reflexes in boys. Although these two studies are well designed and use validated tools in neonates, the precision and predictive validity of neurobehavioral examinations at this early age remain subject to caution, as known from the literature on at-risk populations, such as preterm infants.

Later in infancy, several authors have used the Bayley Scales of Infant Development (BSID), a validated, much translated, and well-published instrument of mental and motor development for children aged 0-42 months. The BSID is currently in its third edition. Kim et al. reported a cohort study, where metabolites of DEHP were measured in the urine of 460 third trimester pregnant women, and 6 months old infants were assessed with the BSID, 2$^{nd}$ edition, which provides a mental (MDI) and psychomotor (PDI) score [44]. There was a statistically significant inverse association between the phthalate levels and the MDI and PDI in boys, after adjusting for potential confounders, and excluding preterm infants. A similar trend was observed for girls, but was not statistically significant. Polanska et al. measured the metabolites of 5 LMWP and 6 HMWP in the urine of 165 third trimester pregnant women and in the urine of their 24 months old children [45]. The children were tested at 24 months with the third edition of the BSID, which gives three composite scores of cognition, language and motor abilities. After adjustments for confounders such as maternal intelligence and socioeconomic status, several prenatal LMWP and HMWP metabolites were inversely correlated with the motor score, but none was associated with cognitive or language scores. There was no significant association between postnatal phthalate levels and the different scores. As part of a larger study, Tellez-Rojo correlated 3$^{rd}$ trimester phthalates with results of the BSID, 2$^{nd}$ edition, at 24, 30 and 36 months [46]. There was no significant association in the whole sample (n = 136), but the girls' scores were inversely associated with DEHP metabolites. Moreover, the boys' motor score was positively associated with three LMWP metabolites. Whyatt et al. measured third trimester DEHP and non-DEHP metabolites and assessed child development at 3 years of age, with the BSID and a standardized questionnaire of child behavior, the Child Behavior Checklist (CBCL) [47]. In the whole sample (n = 296), they found a negative correlation between non-DEHP metabolites and the motor scores. In girls only, there was a negative correlation between non-DEHP metabolites and the mental score and internalizing behavioral problems on the Child Behavior Checklist [62].

For the outcome in middle childhood, we shall first look at the results from studies using formal tests of intelligence, and then at parental questionnaires.

## i) Formal Tests

Recently, Gascon et al. reported on a large prospective follow-up cohort of 367 children for whom DEHP and non-DEHP metabolites were measured in maternal urine at 12 and 32 weeks of gestation, and who were examined at 1, 4 and 7 years old [48]. Mental or cognitive development was assessed at 1 and 4 years old, with the first version of the BSID (1977), and the McCarthy Scales of Children's abilities, a 1972 test. Behavior was assessed at 4 and 7 years old with a social competence scale, a DSM-IV based ADHD questionnaire and two other questionnaires assessing different aspects of behavior. In this study which controlled for many confounders, but used older tests, there was no association between exposure to phthalates and outcomes of developmental tests at 1 or 4 years old. DEHP metabolites were associated with less inattention and ADHD at 4 and 7 years old, and better social competence at 4 years old. There was no difference by trimester of dosage, and no gender effect. The children evaluated at the age of 3 by Whyatt et al. [47] were offered further follow-up at the age of 7 years old (n = 328), with a cognitive test, the Wechsler Intelligence Scales for Children, 4th edition [51]. This test provides a full-scale intelligence quotient (FIQ), a verbal comprehension index, a perceptual index, a processing speed and a working memory index. Furthermore, iterative urinary phthalate levels were measured at the ages of 3 and 5 years old. There were no associations between the children's phthalate levels and any of the cognitive scores. On the other hand, there were negative associations between several maternal non-DEHP metabolites during pregnancy (mono-n-butyl-phthalate and monoisobutyl phthalate) and FIQ, Performance Intelligence Quotient (PIQ), working memory and processing speed, with several differences between genders. The difference in full-scale IQ (Intelligence Quotient) between the lowest and the highest quartile of monoisobutyl phthalate measures for example was 7.6 points (95% CI3.2-12.1), which corresponds to half a standard deviation. These results are in contradiction with Huang et al. who reported the follow-up of a prospective cohort until a maximum of 12 years in 110 children for whom there were urinary maternal phthalate levels, as well as children's urine analysis at follow-up [52]. The children were examined successively at the age of 2-3 years with the BSID, at 5-6 years with the WPPSI-R, and at 8-9 and 11-12 years with the WISC-IV. A mixed-model repeat measure analysis was done, which showed that none of the maternal urinary phthalate metabolites was associated with the cognitive scores, whereas there was *a significant and negative association between DEHP metabolites in the children's urine* and IQ scores. In the 2010 cross-sectional study by Cho et al. DEHP and DBP metabolites in the urine of 621 8-11 year children were negatively associated with full-scale and verbal IQ scores of the short form of the Korean WISC [53].

## ii) Questionnaires

Engel et al. offered further follow-up to the cohort of infants evaluated in the neonatal period with the Brazelton Neonatal Behavioral Assessment Scale [42], and assessed the children at least once (n = 188) between the ages of 4 and 9 years [49]. They used the BRIEF, a 86-item questionnaire encompassing 8 clinical scales of executive function, such as shift, emotional control, working memory, and the Behavior Assessment System for Children-Parent Rating Scales (BASC), a 130-item questionnaire looking at various aspects of behavior, and giving a summary score (Behavior Symptoms Index). They showed that prenatal LMWP metabolites were associated with more aggression, attention and hyperactivity problems, and conduct problems, mainly on the BASC. Chopra et al. published the results of the National Health and Nutrition Examination Survey (NHANES), an annual population based health survey in the United States. The population consisted of 1493 children for whom urinary phthalate metabolites were measured, and for whom the parents reported whether their child had been given a diagnosis of attention deficit disorder (ADD) or of a learning disability [50]. The proportion of children with ADD was high (9.1%) and there was a significant positive association between this diagnosis and the level of DEHP metabolites, as well as with the sum of HMWP.

Several authors have used the Child Behavior Checklist (CBCL), a general questionnaire on behavior. In a multicenter cohort study published in 2014, Kobrosly et al. measured 7 phthalate metabolites in the urine of 153 pregnant women, and assessed child behavior at 6-10 years with the CBCL, controlling for potential confounders [55]. DEHP metabolites were associated with more somatic complaints in boys, whereas DBP metabolites were associated with inattention, aggressiveness, and rule-breaking problems, also in boys. Similarly, Lien et al. published a cohort study (n = 122) examining the association between third trimester DBP and DEHP metabolites and the results of 8-9 year olds on the CBCL [56]. After adjustment, including for child intelligence, both molecules were significantly associated with externalizing, aggressive and delinquent behaviors. However, this is not consistent with the above-mentioned results of Whyatt et al. who found a correlation in younger children with internalizing scores in girls only [47].

Kim et al. reported a cross-sectional study on phthalates and attention deficit and hyperactivity disorder (ADHD) [54]. Four urinary phthalate metabolites (3 DEHP and 1 non DEHP) were measured in 261 school children aged of 8-11 years that were tested with the Continuous performance test (CPT), a computer-based standardized test of attention and impulsivity, and with the short Korean Educational Development Institute-Wechsler Intelligence Scales. The teachers were concurrently asked to fill in the Teachers–Rating ADHD Scale. There was a strong correlation between the children's DEHP metabolite levels and inattention and hyperactivity scores. In a large cross-sectional study of 179 children diagnosed with ADHD, Park et al. correlated CPT results with the children's urine phthalates, and did genotyping for ADHD candidate genes involved in dopaminergic transmission [58]. They found a positive correlation between DEHP metabolites and the number of errors, and moreover, an interaction between genetic subtype and the effect of phthalates on CPT scores, suggesting there might be genetic susceptibility issues which are yet unrecognized. In the same population of children with ADHD, they performed brain Magnetic Resonance Imaging (MRI) (n = 115), and assembled a control group of 438 age and gender matched children for the subgroup of ADHD boys aged 8-11 years old [59]. The phthalate, and particularly the DEHP levels in the ADHD patients were significantly higher than in the controls. Brain MRI showed a negative correlation between DEHP metabolites and cortical thickness in the right middle and superior temporal gyri. In another study, the same group looked also at psychiatric issues, and correlated middle childhood phthalate levels (DEHP and DBP metabolites) with results of the Trait Anxiety Inventory for Children, a standardized questionnaire [57]. They found a negative correlation between DEHP metabolite levels and trait anxiety in girls, while there were no significant associations in boys.

Finally, two case-control studies reported the association of phthalates with the diagnosis of autistic spectrum disorder (ASD) [60, 61]. In the first study, Testa et al. measured four DEHP metabolites in the urine of children diagnosed with ASD (n = 4 5) at an age of 11 +/- 5 years old, after exclusion of patients with known genetic causes of ASD, and in a gender- and age-matched control group (n = 40) [61]. Among the four tested metabolites, three were significantly higher in ASD children. They tested those same metabolites in the urine of children diagnosed with Rett syndrome, a genetic cause of ASD-like behavior, and did not find the same results, suggesting that phthalates are associated with non-genetic ASD.

Philippat et al. measured phthalates in the house dust of 3 groups of children diagnosed with ASD (n = 50), with developmental delay (DD) (n = 27) and in controls (n = 68) [60]. Phthalate levels were associated with the diagnosis of DD, but not ASD. Nevertheless, in ASD and DD children, metabolites levels were positively associated with inattention and hyperactivity symptoms.

In summary, the evidence coming from human epidemiological data about prenatal, and possibly postnatal, exposure to phthalates points in the direction of a global neurodetrimental effect. There is a great heterogeneity between studies, due to different time-points, different molecules, and different outcomes. Reported adverse outcomes concern reduced general cognitive ability, externalizing symptoms with hyperactivity, aggressiveness, disruptive behavior and inattention, mainly in boys, and internalizing behavior and anxiety trait in girls.

## CONCLUSION

Concerns about possible toxic effects of phthalates on the human health have risen since the 1990s, first in consumers, who have encouraged politicians and researchers to finance and design studies in order to assess dangers. Inferring a causal relationship between a potential neurodetrimental agent and a specified outcome is complex, and requires preclinical evidence of brain injury/abnormal development, with a plausible pathophysiological mechanism and a plausible timing relationship between exposure and outcome. Preclinical *in vitro* evidence about phthalates shows detrimental effects on brain development, with a reduction in dendrites and axons, enhanced neuronal death, possibly mediated by apoptosis and inflammation. Animal studies measured mostly visual-spatial learning, memory, and motor activity in rat models and found some impairment. Compatible alterations were found in the corresponding brain region, the hippocampus. In human epidemiological studies, visual-spatial impairments were not prominent. Vocabulary scores and executive functions were mostly affected, which are of course more difficult to assess in animals. The only human imaging study to date showed a thinning in the right temporal cortex, not typically associated with behavior or vocabulary scores.

The current evidence has several limitations; first, extrapolations from animal studies to humans should be subject to great caution. Secondly, epidemiological data are difficult to interpret because notably of the long latency between the exposure and the measurable effects, and of the multifactorial determinants of neurodevelopment, thus potentially allowing for biases and confounders. Furthermore, the limited number of available studies and their strong heterogeneity does not allow yet a clear answer about the effect of phthalates on neurodevelopment.

There is thus a need of large well-designed prospective studies to further investigate long time neurodevelopmental effects of phthalates. Therefore, reliable and easily measurable bio-markers of the exposure remain to be identified.

Current and future challenges for public health will also consist in pursuing the development and study of alternatives to phthalates, especially in the critical field of medical devices. Changes in regulations should thus take into account not only the security concerns about known phthalates, but also the iso-technicity and the uncertainty about the safety of more recent alternatives. New regulations should also include ways to ensure their practical applications.

In summary, the literature so far suggests that perinatal exposure to phthalates may induce alterations in neural growth and differentiation in the brain, with long-lasting consequences on neurodevelopment. The evidence is based on animal and epidemiological studies, and is therefore to be treated with caution. Moreover, one should keep in mind that phthalates are only one of many potential environmental toxics. Nevertheless, there is a growing body of studies, pointing in the same direction and this should encourage decision makers to act on these endocrine, and possibly also neurodevelopmental, disruptors.

## REFERENCES

[1] Jensen M. S., Norgaard-Pedersen B., Toft G., Hougaard D. M., Bonde J. P., Cohen A. et al. Phthalates and perfluorooctanesulfonic acid in human amniotic fluid: temporal trends and timing of amniocentesis in pregnancy. *Environ Health Perspect*. 2012; 120(6):897-903.

[2] Hogberg J., Hanberg A., Berglund M., Skerfving S., Remberger M., Calafat A. M. et al. Phthalate diesters and their metabolites in human breast milk, blood or serum, and urine as biomarkers of exposure in vulnerable populations. *Environ Health Perspect*. 2008; 116(3):334-9.

[3] IARC. *Di (2-ethylhexyl) phthalate (DEHP)*. International Agency for Research on Cancer; 2000.

[4] Fischer C. J., Bickle Graz M., Muehlethaler V., Palmero D., Tolsa J. F. Phthalates in the NICU: is it safe? *J Paediatr Child Health*. 2013; 49(9): E413-9.

[5] Fischer Fumeaux C. J., Stadelmann Diaw C., Palmero D., M'Madi F., Tolsa J. F. [Phthalate exposure in the neonatal intensive care unit]. *Archives de pediatrie: organe officiel de la Societe francaise de pediatrie*. 2015; 22(2):203-10.

[6] Goodman M., LaKind J. S., Mattison D. R. Do phthalates act as obesogens in humans? A systematic review of the epidemiological literature. *Critical Reviews in Toxicology*. 2014; 44(2):151-75.

[7] Yaghjyan L., Sites S., Ruan Y., Chang S. H. Associations of urinary phthalates with body mass index, waist circumference and serum lipids among females: National Health and Nutrition Examination Survey 1999-2004. *International journal of obesity*. 2015; 39(6):994-1000.

[8] Albert O., Jegou B. A critical assessment of the endocrine susceptibility of the human testis to phthalates from fetal life to adulthood. *Hum Reprod Update*. 2014; 20(2):231-49.

[9] Bornehag C. G., Carlstedt F., Jonsson B. A., Lindh C. H., Jensen T. K., Bodin A. et al. Prenatal phthalate exposures and anogenital distance in Swedish boys. *Environ Health Perspect*. 2015; 123(1):101-7.

[10] Swan S. H., Sathyanarayana S., Barrett E. S., Janssen S., Liu F., Nguyen R. H. et al. First trimester phthalate exposure and anogenital distance in newborns. *Hum Reprod*. 2015; 30(4):963-72.

[11] Marie C., Vendittelli F., Sauvant-Rochat M.-P. Obstetrical outcomes and biomarkers to assess exposure to phthalates: A review. *Environment International*. 2015; 83:116-36.

[12] Ku H. Y., Su P. H., Wen H. J., Sun H. L., Wang C. J., Chen H. Y. et al. Prenatal and postnatal exposure to phthalate esters and asthma: a 9-year follow-up study of a taiwanese birth cohort. *PLoS One*. 2015; 10(4): e0123309.

[13] Wang I. J., Lin C. C., Lin Y. J., Hsieh W. S., Chen P. C. Early life phthalate exposure and atopic disorders in children: a prospective birth cohort study. *Environ Int*. 2014; 62:48-54.

[14] Kim S. H., Park M. J. Phthalate exposure and childhood obesity. *Annals of Pediatric Endocrinology & Metabolism*. 2014; 19(2):69-75.

[15] Trasande L., Sathyanarayana S., Spanier A. J., Trachtman H., Attina T. M., Urbina E. M. Urinary phthalates are associated with higher blood pressure in childhood. *J. Pediatr*. 2013; 163(3):747-53 e1.

[16] Trasande L., Attina T. M. Association of Exposure to Di-2-Ethylhexylphthalate Replacements with Increased Blood Pressure in Children and Adolescents. *Hypertension*. 2015; 66(2):301-8.

[17] Ejaredar M., Nyanza E. C., Ten Eycke K., Dewey D. Phthalate exposure and childrens neurodevelopment: A systematic review. *Environ Res*. 2015; 142:51-60.

[18] Miodovnik A., Edwards A., Bellinger D. C., Hauser R. Developmental neurotoxicity of ortho-phthalate diesters: review of human and experimental evidence. *Neurotoxicology*. 2014; 41:112-22.

[19] Kasuya M. Toxicity of phthalate esters to nervous tissue in culture. *Bulletin of environmental contamination and toxicology*. 1974; 12(2): 167-72.

[20] Lee T. W., Tumanov S., Villas-Boas S. G., Montgomery J. M., Birch N. P. Chemicals eluting from disposable plastic syringes and syringe filters alter neurite growth, axogenesis and the microtubule cytoskeleton in

cultured hippocampal neurons. *Journal of neurochemistry*. 2015; 133(1): 53-65.

[21] Chen T., Yang W., Li Y., Chen X., Xu S. Mono-(2-ethylhexyl) phthalate impairs neurodevelopment: inhibition of proliferation and promotion of differentiation in PC12 cells. *Toxicology letters*. 2011; 201(1):34-41.

[22] Aung K. H., Win-Shwe T. T., Kanaya M., Takano H., Tsukahara S. Involvement of hemeoxygenase-1 in di(2-ethylhexyl) phthalate (DEHP)-induced apoptosis of Neuro-2a cells. *The Journal of toxicological sciences*. 2014; 39(2):217-29.

[23] Guida N., Laudati G., Galgani M., Santopaolo M., Montuori P., Triassi M, et al. Histone deacetylase 4 promotes ubiquitin-dependent proteasomal degradation of Sp3 in SH-SY5Y cells treated with di(2-ethylhexyl) phthalate (DEHP), determining neuronal death. *Toxicol Appl Pharmacol*. 2014; 280(1):190-8.

[24] Ishido M., Suzuki J. Classification of phthalates based on an *in vitro* neurosphere assay using rat mesencephalic neural stem cells. *The Journal of toxicological sciences*. 2014; 39(1):25-32.

[25] Wu Y., Li K., Zuo H., Yuan Y., Sun Y, Yang X. Primary neuronal-astrocytic co-culture platform for neurotoxicity assessment of di-(2-ethylhexyl) phthalate. *Journal of environmental sciences*. 2014; 26(5): 1145-53.

[26] Clancy B., Finlay B. L., Darlington R. B., Anand K. J. Extrapolating brain development from experimental species to humans. *Neurotoxicology*. 2007; 28(5):931-7.

[27] Workman A. D., Charvet C. J., Clancy B., Darlington R. B., Finlay B. L. Modeling transformations of neurodevelopmental sequences across mammalian species. *The Journal of neuroscience: the official journal of the Society for Neuroscience*. 2013; 33(17):7368-83.

[28] Andrade A. J. M., Grande S. W., Talsness C. E., Gericke C., Grote K., Golombiewski A. et al. A dose response study following in utero and lactational exposure to di-(2-ethylhexyl) phthalate (DEHP): Reproductive effects on adult male offspring rats. *Toxicology*. 2006; 228(1):85-97.

[29] Dabholkar A. S. Peroxisomes in the rat brain and the effects of di-(2-ethylhexyl) phthalate during postnatal development. An electron-microscopic study. *Acta anatomica*. 1988; 131(3):218-21.

[30] Ishido M., Masuo Y., Sayato-Suzuki J., Oka S., Niki E., Morita M. Dicyclohexylphthalate causes hyperactivity in the rat concomitantly with

impairment of tyrosine hydroxylase immunoreactivity. *Journal of neurochemistry*. 2004; 91(1):69-76.

[31] Li X., Jiang L., Cheng L., Chen H. Dibutyl phthalate-induced neurotoxicity in the brain of immature and mature rat offspring. *Brain & development*. 2014; 36(8):653-60.

[32] Li X.-J., Jiang L., Chen L., Chen H.-S., Li X. Neurotoxicity of dibutyl phthalate in brain development following perinatal exposure: A study in rats. *Environmental toxicology and pharmacology*. 2013; 36(2):392-402.

[33] Li Y., Li T., Zhuang M., Wang K., Zhang J., Shi N. High-dose dibutyl phthalate improves performance of F1 generation male rats in spatial learning and increases hippocampal BDNF expression independent on p-CREB immunocontent. *Environmental toxicology and pharmacology*. 2010; 29(1):32-8.

[34] Min A., Liu F., Yang X., Chen M. Benzyl butyl phthalate exposure impairs learning and memory and attenuates neurotransmission and CREB phosphorylation in mice. *Food and chemical toxicology: an international journal published for the British Industrial Biological Research Association*. 2014; 71:81-9.

[35] Peng L. Mice Brain Tissue Injury Induced by Diisononyl Phthalate Exposure and the Protective Application of Vitamin E. *Journal of biochemical and molecular toxicology*. 2015; 29(7):311-20.

[36] Smith C. A., Holahan M. R. Reduced hippocampal dendritic spine density and BDNF expression following acute postnatal exposure to di(2-ethylhexyl) phthalate in male Long Evans rats. *PLoS One*. 2014; 9 (10):e109522.

[37] Smith C., MacDonald A., Holahan M. R. Acute postnatal exposure to di(2-ethylhexyl) phthalate adversely impacts hippocampal development in the male rat. *Neuroscience*. 2011; 193(0):100-8.

[38] Sun W., Ban J.-B., Zhang N., Zu Y.-K., Sun W.-X. Perinatal exposure to di-(2-ethylhexyl)-phthalate leads to cognitive dysfunction and phospho-tau level increase in aged rats. *Environmental toxicology*. 2012; 29(5): 596-603.

[39] Tanaka T. Reproductive and neurobehavioural toxicity study of bis(2-ethylhexyl) phthalate (DEHP) administered to mice in the diet. *Food and chemical toxicology: an international journal published for the British Industrial Biological Research Association*. 2002; 40(10):1499-506.

[40] Li X. J., Jiang L., Chen L., Chen H. S., Li X. Neurotoxicity of dibutyl phthalate in brain development following perinatal exposure: a study in rats. *Environmental toxicology and pharmacology*. 2013; 36(2):392-402.

[41] Tanaka T. Reproductive and neurobehavioural effects of bis(2-ethylhexyl) phthalate (DEHP) in a cross-mating toxicity study of mice. *Food and chemical toxicology: an international journal published for the British Industrial Biological Research Association*. 2005; 43(4): 581-9.

[42] Engel S. M., Zhu C., Berkowitz G. S., Calafat A. M., Silva M. J., Miodovnik A. et al. Prenatal phthalate exposure and performance on the Neonatal Behavioral Assessment Scale in a multiethnic birth cohort. *Neurotoxicology*. 2009; 30(4):522-8.

[43] Yolton K., Xu Y., Strauss D., Altaye M., Calafat A. M., Khoury J. Prenatal exposure to bisphenol A and phthalates and infant neurobehavior. *Neurotoxicology and Teratology*. 2011; 33(5):558-66.

[44] Kim Y., Ha E. H., Kim E. J., Park H., Ha M., Kim J. H. et al. Prenatal exposure to phthalates and infant development at 6 months: prospective Mothers and Children's Environmental Health (MOCEH) study. *Environ Health Perspect*. 2011; 119(10):1495-500.

[45] Polanska K., Ligocka D., Sobala W., Hanke W. Phthalate exposure and child development: the Polish Mother and Child Cohort Study. *Early Hum Dev*. 2014;90(9):477-85.

[46] Tellez-Rojo M. M., Cantoral A., Cantonwine D. E., Schnaas L., Peterson K., Hu H. et al. Prenatal urinary phthalate metabolites levels and neurodevelopment in children at two and three years of age. *Science of the Total Environment*. 2013; 461-462(0):386-90.

[47] Whyatt R. M., Liu X., Rauh V. A., Calafat A. M., Just A. C., Hoepner L. et al. Maternal Prenatal Urinary Phthalate Metabolite Concentrations and Child Mental, Psychomotor, and Behavioral Development at 3 Years of Age. *Environ Health Perspect*. 2012; 120(2):290-5.

[48] Gascon M., Valvi D., Forns J., Casas M., Martínez D., Júlvez J. et al. Prenatal exposure to phthalates and neuropsychological development during childhood. *International Journal of Hygiene and Environmental Health*. 2015; 218(6):550-8.

[49] Engel S. M., Miodovnik A., Canfield R. L., Zhu C., Silva M. J., Calafat A. M. et al. Prenatal phthalate exposure is associated with childhood behavior and executive functioning. *Environ Health Perspect*. 2010; 118(4):565-71.

[50] Chopra V., Harley K., Lahiff M., Eskenazi B. Association between phthalates and attention deficit disorder and learning disability in U.S. children, 6-15 years. *Environ Res*. 2014;128:64-9.

[51] Factor-Litvak P., Insel B., Calafat A. M., Liu X., Perera F., Rauh V. A. et al. Persistent Associations between Maternal Prenatal Exposure to Phthalates on Child IQ at Age 7 Years. *PLoS ONE*. 2014; 9(12):e114 003.

[52] Huang H. B., Chen H. Y., Su P. H., Huang P. C., Sun C. W., Wang C. J., et al. Fetal and Childhood Exposure to Phthalate Diesters and Cognitive Function in Children Up to 12 Years of Age: Taiwanese Maternal and Infant Cohort Study. *PLoS One*. 2015; 10(6):e0131910.

[53] Cho S. C., Bhang S. Y., Hong Y. C., Shin M. S., Kim B. N., Kim J. W. et al. Relationship between environmental phthalate exposure and the intelligence of school-age children. *Environ Health Perspect*. 2010; 118(7):1027-32.

[54] Kim B. N., Cho S. C., Kim Y., Shin M. S., Yoo H. J., Kim J. W. et al. Phthalates exposure and attention-deficit/hyperactivity disorder in school-age children. *Biological psychiatry*. 2009; 66(10):958-63.

[55] Kobrosly R. W., Evans S., Miodovnik A., Barrett E. S., Thurston S. W., Calafat A. M. et al. Prenatal phthalate exposures and neurobehavioral development scores in boys and girls at 6-10 years of age. *Environ Health Perspect*. 2014; 122(5):521-8.

[56] Lien Y. J., Ku H. Y., Su P. H., Chen S. J., Chen H. Y., Liao P. C. et al. Prenatal exposure to phthalate esters and behavioral syndromes in children at 8 years of age: Taiwan Maternal and Infant Cohort Study. *Environ Health Perspect*. 2015; 123(1):95-100.

[57] Park S., Cheong J. H., Cho S. C., Kim J. W., Shin M. S., Yoo H. J. et al. Di-(2-ethylhexyl) phthalate exposure is negatively correlated with trait anxiety in girls but not with trait anxiety in boys or anxiety-like behavior in male mice. *J Child Neurol*. 2015; 30(1):48-52.

[58] Park S., Kim B. N., Cho S. C., Kim Y., Kim J. W., Lee J. Y. et al. Association between urine phthalate levels and poor attentional performance in children with attention-deficit hyperactivity disorder with evidence of dopamine gene-phthalate interaction. *International journal of environmental research and public health*. 2014; 11(7):6743-56.

[59] Park S., Lee J. M., Kim J. W., Cheong J. H., Yun H. J., Hong Y. C. et al. Association between phthalates and externalizing behaviors and cortical thickness in children with attention deficit hyperactivity disorder. *Psychological medicine*. 2014:1-12.

[60] Philippat C., Bennett D. H., Krakowiak P., Rose M., Hwang H. M., Hertz-Picciotto I. Phthalate concentrations in house dust in relation to

autism spectrum disorder and developmental delay in the CHildhood Autism Risks from Genetics and the Environment (CHARGE) study. *Environmental health: a global access science source*. 2015; 14:56.

[61] Testa C. N. F., Hayek J., De Felice C., Chelli M., Rovero P., Latini G., Papini A. M. Di(2-ethylhexyl)phthalate and Autism Spectrum Disorders. *ASN Neuro*. 2012; 4(Apr 27).

[62] Achenbach T. M., Ruffle T. M. The Child Behavior Checklist and related forms for assessing behavioral/emotional problems and competencies. *Pediatrics in review/American Academy of Pediatrics*. 2000; 21(8):265-71.

In: Phthalates
Editor: Miguel Owens
ISBN: 978-1-63483-908-2


*Chapter 2*

# A Critical Review on the Chromatographic Detection of Phthalates in Different Matrices: Problems and Solutions

***Mario Vincenzo Russo*[1,*] *and Pasquale Avino*[2]**

[1]Dept. Agriculture, Environment and Food,
University of Molise, Campobasso, Italy
[2]DIT, INAIL Research Area, Rome, Italy

## Abstract

Phthalates, chemicals easily released into the environment, are found to be ubiquitous. Although some differences in toxicity are well established, current data are not sufficient to allow elucidation of the range of toxicological differences. One of the most important issues for these compounds regards the analytical determination: many papers in literature can be found using chromatographic methods. This paper would like to show and investigate the advantages and/or the disadvantages of some analytical methods used these last 5 years. More than 150 papers have been highlighted and a critical revision is performed in relationship to complex matrices such as food, beverage and vegetables.

---

* Corresponding author: Dept. Agriculture, Environment and Food, University of Molise, via De Sanctis, I-86010 Campobasso (Italy). Ph.: +39 0874 404 631; Fax: +39 0874 404 652; E-mail: mvrusso@unimol.it.

**Keywords:** phthalates, analysis, determination, food, beverage, pasta, rice, meat, fish, vegetable, egg, cheese, fruit, coffee, tea, cigarette, wine, alcohol, milk, drink

## INTRODUCTION

Phthalates (Table 1) are chemical plasticizers that have been widely used since the 1950s to soften plastics that would otherwise be brittle and crack when bent (Fromme et al. 2002): nowadays, they are largely used in polymers to enhance the flexibility and extensibility of the materials. From a chemical point of view phthalates are compounds primarily used to make polyvinyl chloride (PVC) or vinyl flexible and pliant (Bart 2006): for this reason they are plasticizers often used in the production of many types of plastics, certain inks, paints, and other products. Further, they are also used in hundreds of products in our homes, hospitals, cars and businesses (US ATSDR 1995; US ATSDR 1997a; US ATSDR 1997b; US ATSDR 2002; US EPA, 2002; US NTP 2007). Their large success is due to their strong performance, durability and stability (Crossfire Beistein 2009; HSDB 2009). Because these phthalates plasticizers are bound into the material in which they are added, they do not easily migrate out of the product or evaporate (Clara et al. 2010; Hubinger. 2010).

Phthalates are categorized as high and low, depending on their molecular weight. In particular, high phthalates are those with 7-13 Carbon atoms in their chemical backbone, which gives them increased permanency and durability. The most common types of high phthalates include diisononyl phthalate (DINP), diisodecyl phthalate (DIDP) and dipropylheptyl phthalate (DPHP). High phthalates are commonly used in PVC products such as wire and cable, flooring, wall covering, self-adhesive films, synthetic leather, coated fabrics and roofing and automobile applications (Ash and Ash 1987). On the other hand, low phthalates include those with 3-6 carbon atoms in their chemical backbone (i.e., in the branched chain). The most common types of low phthalates include di(2-ethylhexyl) phthalate (DEHP) and dibutyl phthalate (DBP) (Giust 1990; Sharman et al. 1994). Low phthalates are commonly used in medical devices, general purpose PVC, adhesives and inks (Hillman et al. 1975; Roth et al. 1988; Christensson et al. 1991; Tickner et al. 2001; Hansen 2007; Van Vliet et al. 2011).

Worldwide production of PAEs is approximately 6 million tons per year (Mackintosh et al. 2006): with a wide range of physical and chemical properties, phthalates are used in a multitude of consumer and industrial products that demand high performance, long-lasting wear and durability. They can be employed in a variety of applications even if the phthalates are not necessarily interchangeable. The characteristics of an individual phthalate often make it well suited to a particular product, allowing manufacturers to meet unique requirements for its use (function and safety specifications), appearance (texture, color, size and shape), and durability and wear. For this reason, substitutions could sacrifice the functionality, quality, longevity, cost or performance of a product.

Plasticizers are worldwide distributed; in particular children have many opportunities to absorb them (Karle et al. 1997; Oie et al. 1997; Loff et al. 2000; Kambia et al. 2001). As reported by Sathyanarayana et al. (2008), children are uniquely vulnerable to phthalate exposures given their hand-to-mouth behaviors, floor play, and developing nervous and reproductive systems. Similar studies are also carried out by other authors (Sjoberg et al. 1985a; Sjoberg et al. 1985b; Steiner et al. 1998; Wilkinson and Lamb 1999; Petersen and Breindahl 2000; Stringer et al. 2000; Bouma and Schakel. 2002; Shea 2003).

In fact, because phthalates are not chemically bound to the plastics they are added to, they are continuously released into the air or food or liquid (Bosnir et al. 2003; Teil et al. 2006; Cao 2010). With respect to health effects, phthalates are often classified as endocrine disruptors or hormonally-active agents (HAAs) because of their ability to interfere with the endocrine system in the body (Vitali et al. 1997; Latini 2005; Hauser and Calafat 2005; Schettler 2006; Latini et al. 2006; US NTP 2007; Wigle et al. 2008).

The American Congress has permanently banned three types of phthalates (di-(2-ethylhexyl) phthalate DEHP, dibutyl phthalate DBP, or benzyl butyl phthalate BBP) in any amount greater than 0.1% (computed for each phthalate, individually) in children's toys and certain child care articles. The authors would like to remember that a children's toy is defined as a consumer product designed or intended by the manufacturer for a child who is 12 years old or younger for use by the child when the child plays.

Basically, exposure to phthalate esters occurs through the air we breathe, as well as our daily contact with consumer products and food, including beverages. There have been reports from the U.S. Food and Drug Administration (US FDA) that certain foods and beverages, particularly fruit juices, contain high levels of phthalates. In some cases, deliberate adulteration of soft drinks with phthalate esters has been reported (Mervish et al. 2014; Law 2015).

Because these compounds are reported to act as endocrine disruptors, and exposure to high levels can cause harmful effects in the human reproductive system, their determination in food and beverages are fundamental. The PAEs levels in foods are highly variable and suggest contamination events which may occur in the source or original collection of the food, in the packaging or processing prior to market, or in the preparation leading to consumption (Rodgers et al. 2014). The migration of chemicals from paper and/or board into dry foodstuffs is being extensively studied (Vera et al. 2011; Aznar et al. 2011). For instance, diisopropylnaphthalenes (DIPN) have been found to migrate into rice (Sturaro et al. 1994), into rice, pasta and flour (Boccacci Mariani et al. 1999) and, along with dibutylphthalate (DiBP), into rice, flour and icing sugar (Summerfield and Cooper 2001). PAEs are also found to migrate from paper into sugar (Aurela et al. 1999).

In this review we point out a critical exam on the analytical methods available in literature for determining PAEs in food matrices. In particular, based on Scopus data-base (using phthalate and analysis and matrix name as keywords), we investigated the recent literature (from 2010) on the analytical PAE determination in pasta, rice, meat, fish, cheese, egg, vegetable, fruit, coffee, tea, beverages, milk and cigarettes. Very few studies are present for pasta and rice (2 and 6 papers, respectively) whereas meat, fish and vegetable are well-studied matrices (19, 48 and 46 papers, respectively); about beverages, 44 papers investigate the PAE determination in wine and high alcoholic drink and soft drinks including also fruit juices or other baby beverages whereas 30 papers regard the PAE analysis in mammalian or commercial milk. Of course, in some cases an analytical protocol reported in a paper is validated, and valid as well, for more than one matrix: for instance, the very interesting study by Ji et al. (2014) shows the PAE determination (along with the investigation of analytical parameters) in about 450 samples including foodstuffs (rice, steamed bun, vegetables, meat, poultry, fish, milk and fruits), ambient PM10, drinking water, soil, indoor PM10 and indoor dust samples. On the other hand, there are papers reporting the PAE content without any analytical information. On view of this, the authors have tried to review the best analytical performances for each matrix.

## PHTHALATE DETERMINATION IN FOOD AND BEVERAGES

Phthalates, especially DEHP, have been reported at high concentrations in some foods. Additional evidence that food and food packaging are important sources of DEHP exposure come from observational human epidemiologic and panel studies.

For these reasons, we extensively analysed the analytical methods along with their advantages for PAEs analysis in each alimentary matrix (Table 2): about that, we preferred to divide the approaches according the various foods and beverages. In general, gas-chromatography (GC) or liquid chromatography (LC) analysis followed by mass spectrometry (MS) detection are used for PAE evaluation: the clean-up step depends on the different matrix analyzed.

We would like to remember that the Limit of Detection (LOD) for chromatographic separations typically corresponds to the concentration peak with signal-to-noise (S/N) of 3 (Knoll 1985; Armbruster et al. 1994; Dolan 2006; Boqué and Heyden 2009) whereas the Limit of Quantification (LOQ) to the concentration peak with S/N of 10 (Armbruster et al. 1994; Armbruster and Pry 2008). The S/N definition for chromatography is the ratio of the peak height relative to the middle of the noise range (S) to the difference between the maximum and minimum baseline signal values for the noise (N) (Dolan 2006).

## Pasta and Rice Matrices

For these two matrices (Table 2) in literature there is present just one paper dealing the PAE determination, Purcaro et al. (2013). The authors focused their attention on the development/optimization of a comprehensive two-dimensional gas chromatography method, with flame ionization (FID) and mass spectrometric detection, for the simultaneous identification and quantification of mineral-oil contaminants in a variety of food products. Two main classes of contaminants, namely saturated and aromatic hydrocarbons and including di(ethylhexyl) phthalate, were previously fractionated on a manually-packed silver silica solid-phase extraction (SPE) cartridge. The authors applied this procedure to different food products (pasta, rice and icing sugar): they investigated the analytical determination in 0.35-24 mg $kg^{-1}$ range, reaching a very good coefficient of regression (0.9993) and a significant coefficient of variation (CV%) of the slope (1%), and a valuable Limit of Quantification (LOQ) (1.2 mg $kg^{-1}$).

**Table 1. Main Phthalates (PAEs) diffused worldwide, with their corresponding abbreviations, chemical formulas and CAS numbers**

| Phthalate | Abbreviation | Formula | CAS number |
|---|---|---|---|
| Di-methyl phthalate | DMP | $C_{10}H_{10}O_4$ | 131-113 |
| Di-ethyl phthalate | DEP | $C_{12}H_{14}O_4$ | 84-66-2 |
| Di-allyl phthalate | DAP | $C_{14}H_{14}O_4$ | 131-17-9 |
| Di-*n*-propyl phthalate | DPP | $C_{14}H_{18}O_4$ | 131-16-8 |
| Di-*n*-butyl phthalate | DBP | $C_{16}H_{22}O_4$ | 84-74-2 |
| Di-isobutyl phthalate | DiBP | $C_{16}H_{22}O_4$ | 84-69-5 |
| Butyl cyclohexyl phthalate | BCP | $C_{18}H_{24}O_4$ | 84-64-0 |
| Di-*n*-pentyl phthalate | DnPP | $C_{18}H_{26}O_4$ | 131-18-0 |
| Di-cyclohexyl phthalate | DCP | $C_{20}H_{26}O_4$ | 84-61-7 |
| Butyl benzyl phthalate | BBP | $C_{19}H_{20}O_4$ | 85-68-7 |
| Di-*n*-hexyl phthalate | DnHP | $C_{20}H_{30}O_4$ | 84-75-3 |
| Di-isohexyl phthalate | DiHxP | $C_{20}H_{30}O_4$ | 146-50-9 |
| Di-isoheptyl phthalate | DiHpP | $C_{22}H_{34}O_4$ | 41451-28-0 |
| Butyl decyl phthalate | BDP | $C_{22}H_{34}O_4$ | 89-19-0 |
| Bis-(2-ethylhexyl) phthalate | DEHP, DOP | $C_{24}H_{38}O_4$ | 117-81-7 |
| Di-(*n*-octyl) phthalate | DnOP | $C_{24}H_{38}O_4$ | 117-84-0 |
| Di-isooctyl phthalate | DiOP | $C_{24}H_{38}O_4$ | 27554-26-3 |
| *n*-Octyl *n*-decyl phthalate | ODP | $C_{26}H_{42}O_4$ | 119-07-3 |
| Di-isononyl phthalate | DiNP | $C_{26}H_{42}O_4$ | 28553-12-0 |
| Di-(2-propyltheptyl) phthalate | DPHP | $C_{28}H_{46}O_4$ | 53306-54-0 |
| Di-isodecyl phthalate | DiDP | $C_{28}H_{46}O_4$ | 26761-40-0 |
| Di-undecyl phthalate | DUP | $C_{30}H_{50}O_4$ | 3648-20-2 |
| Di-isoundecyl phthalate | DiUP | $C_{30}H_{50}O_4$ | 85507-79-5 |
| Di-tridecyl phthalate | DtDP | $C_{34}H_{58}O_4$ | 119-06-2 |
| Di-isotridecyl phthalate | DiUP | $C_{34}H_{58}O_4$ | 68515-47-9 |

## Meat and Fish Matrices

As aspects regarding migration of phthalates from packaging films into foods, fatty and oily ones are primarily contaminated by phthalates due to their lipophilicity. For this reason, the meat matrix is strongly exposed to the possible PAE contamination. During these last 5 years different papers have been published in this field (Table 2). Similar to previous and further matrices, some papers are devoted to the PAE evaluation in some diets or subpopulation without main information on the analytical paramenters (Colacino et al. 2010;

Kappenstein et al. 2012; Sioen et al. 2012; Bradley et al. 2013; Trasande et al. 2013; Mervish et al. 2014; Fierens et al. 2014; Kim et al. 2014).

A method for the determination of six phthalates (DMP, DEP, DBP, BBP, DEHP and DOP) in ham sausage samples utilizing a comprehensive sample pre-treatment procedure, including liquid extraction with n-hexane and solid-phase extraction (SPE) with Oasis MAX cartridges, in combination with GC-MS was developed by Guo et al. (2010). The linearity range of the method is tested between 0.5-1000 ng $g^{-1}$ with all correlation coefficients greater than 0.99. The results of the percentage recovery and coefficient of variation (CV%) show that the method is accurate and precise (recoveries from 90.8% to 106.2% for all phthalates with a CV value from 3.7% to 8.2%) and LODs and LOQs range between 0.29-0.61 ng $g^{-1}$ and 0.53-2.01 ng $g^{-1}$, respectively.

Cao et al. (2014) investigated the occurrence of di-(2-ethylhexyl) adipate and PAE (DEP, DiBP, DBP, BBzP, DEHA, DEHP, DOP) plasticizers in samples of meat, fish, and cheese and their packaging films by extraction with acetonitrile and analysis by GC-MS. Such protocol allows the authors to reach good correlation coefficients (> 0.999) in the concentration range 0.02-10 ng $mL^{-1}$ with MDL between 0.0016-0.43 μg $g^{-1}$ recoveries between 92-108%.

A rapid, sensitive, and accurate method for determining 23 phthalates in foodstuffs including milk-based products, distilled liquor, wine, beverage, grain, meat, oil, biscuit (cookie), and canned food, was developed by Xu et al. (2014). Liquid samples are exacted by acetonitrile, whereas solid samples are prepared by vortexing after addition of acetonitrile and $MgSO_4$ or glass-based SPE methods; the analysis is performed by LC-MS/MS. The LODs are between 0.8 and 15 ng $g^{-1}$ and LOQs were between 10 and 100 ng $g^{-1}$; the recoveries are within 75.5-115.2% and the relative standard deviations (RSDs) are in the range of 3.2-18.9%. Using this protocol the authors analyzed 623 real samples collected in Taiwan, the DEHP was found in almost all samples with levels ranging between 0.02-2685 mg $kg^{-1}$.

An interesting application of the analytical method developed by Fierens and coauthors (Fierens et al. 2012a, 2012b) (which will be described in next sections) is reported in Sakhi et al. (2014): the concentration levels of ten different phthalates and bisphenol A (BPA) in foods and beverages purchased on the Norwegian market are extensively reported and discussed along with the estimation of the daily dietary exposure in the Norwegian adult population. The analytical protocol foresees an extraction with dichloromethane and GC-MS analysis; furthermore, for high fat and some low-fat samples, an extra fat removing purification step is performed by gel permeation chromatography (GPC). In general, PAE and BPA levels in foods and beverages from the

Norwegian market are comparable to other countries worldwide. Grain and meat products are the major contributors of exposure to these chemicals in the Norwegian adult population.

Finally, it should be cited an innovative and powerful sample preparation technique, which is quantitative, simple, environmentally friendly, and economical, and it has high sensitivity and good reproducibility for the simultaneous GC-MS analysis of PAEs from foodstuffs (He et al. 2015). The paper proposes a way to overcome the largest problem in PAEs analysis, i.e., the high blank value because PAEs are widely used in various applications and products. To do this shortcoming, gas purge microsyringe extraction (GP-MSE) is applied, which established a new and low-blank-value analytical method for PAE analysis to analyse PAEs in foodstuffs. Before the GP-MSE procedure, the compounds are ultrasonically extracted with dichloromethane, followed by centrifugation. The supernatant (dichloromethane layer) was transferred to a clean glass bottle. Finally, a GC-MS analysis is performed for the determination. In this study, GP-MSE is used as a clean-up method, and the overall recoveries ranged from 86 to 102%, and the RSD is less than 10%. Further, the results show that the LODs range from 0.14 to 0.38 ng $g^{-1}$ for solid sample and from 0.0021 to 0.0096 ng $mL^{-1}$ for liquid sample whereas the linearity of the method, has been evaluated in the concentration range 5-1500 ng $mL^{-1}$ showing good linearity ($R^2 > 0.999$) for all analytes. The authors applied this method to the PAE determination in 78 foodstuffs divided into six categories, cereal products, snacks, beverages, condiments, seafood, and meat products.

Owing to their lipophilic nature, PAEs are released mainly into foodstuffs containing fat. In this way, fish is a matrix suffering of this contamination because the packaging is really important, and consequently the PAE determination is an essential point. Different papers regard a prospective of the fish consumption and the relative human health, expecially according to the diet or feeding (Fatoki et al. 2010; Wu et al. 2010; Carnevali et al. 2010; Manilal et al. 2011; Sánchez-Avila et al. 2012) or sub-population (Tyrrell et al. 2013; Trasande et al. 2013; Bizarro et al. 2014; Mervish et al. 2014; Chen et al. 2015). On the other hand, more than 10 papers deal the PAE determination in fish (raw) food, also describing the analytical procedure. Jarošová et al. (2010) report a method based on feed samples extracted by a mixture of diluents (hexane:acetone, 1:1); particularly, PAEs are separated from coextracts by gel permeation chromatography on gel Bio beads. For cleaning the compound feed eluates, a cleaning procedure with sulphuric acid is applied. Detection and quantification of PAEs were performed by LC with UV and MS detection. By this protocol they reach LODs of 0.03 μg $g^{-1}$ of DBP and DEHP for animal and vegetable material with low fat contents.

Ostrovský et al. (2011) set up a development of a single GC method for determination of four phthalates (DEP, DBP, DEHP and DDP) in fatty matrices. The authors suggest that the method could be applied as preliminary step of quality control for assessing total contamination with PAEs before a more detailed and targeted PAEs analysis. The method is based on alkaline hydrolysis (at 80 °C for 20 h) of all phthalates to phthalic acid (PA) followed by selective removal of lipophilic substances at pH 1 with n-hexane and derivatization of PA to DMP with diazomethane in aqueous-chloroform; the resulting DMP is absorbed in chloroform and determined by GC-FID method. The method calibration results linear ($R^2$ 0.9966) in the range 1-500 μg $mL^{-1}$ and shows LOD and LOQ of 0.4 and 1.2 μg $g^{-1}$ DMP, respectively.

Adeniyi et al. (2011) detected and quantified the levels of four PAEs (DMP, DEP, DBP, DEHP) in water, sediment, and some fish species using dichloromethane as extraction solvent and a GC-Flame Ionization Detector (GC-FID) for the analysis. The PAEs are investigate in the linear working range 2-4000 μg $L^{-1}$ showing a $R^2$ 0.99 for all the PAEs. The Method Detection Limit (MDL) (Miller and Miller 1998) and the instrument detection limit values are (in μg $L^{-1}$): DMP, 40.17 and 20.46; DEP, 48.89 and 10.51; DBP, 87.52 and 3.80; DEHP, 14.61 and 1.09, respectively. The percentage PAE recoveries range between 82% and 91%. Using this protocol the authors analyzed twenty-four water, sediment, and fish samples: the main findings are that i) DMP is not detected in the water and fish samples but is detected in sediments collected from four of the six sampling sites, and ii) the differences in fish PAE levels are not statistically significant, meaning PAEs accumulation to be not fish species dependent.

One of the most active researchers in the PAE field is Fierens. Among different papers published, a very original and interesting study regards the PAE contamination generated in foods or the chemicals migration from food contact materials into foods during processing (Fieresens et al. 2012a). In particular, the authors examine the effect of cooking at home on the levels of PAEs (DMP, DEP, DiBP, DnBP, BBP, DEHP, DCHP, DnOP) in various food types (starchy products, vegetables and meat and fish). Food products are analyzed before as well as after cooking (boiling, steaming, (deep-)frying or grilling): they find a PAE content decline in foods after cooking, except in vegetables, where no relevant effect is noted. They report that several factors influence the degree of this decline (e.g. weight difference, fat uptake, etc.), concluding that cooking at home definitely affects the PAE concentrations in foods and thus needs to be considered in order to correctly assess humans' dietary exposure to such contaminants. Due to their interest for the PAE content during processing food, no many analytical parameters are reported except some generic information regarding the well reproducible analytical procedure as recoveries, varying between 93% and 100% for high-fat foods and between 88% and 102% for low-fat foods, the RSDs and measurement uncertainties lower than 11% and 30% for each phthalate compound, respectively.

Bradley et al. (2013a, 2013b) developed a method for determining different PAEs (DMP, DEP, DAP, BBP, DCHP, DHexP, DEHP, DiNP, DiPP, DiBP, DiDP, DPP, DBP, DOP, DDP) in different foods including fish. The method was validated in-house and its broad applicability demonstrated by the analysis of high-fat, high-carbohydrate and high-protein foodstuffs as well as combinations of all three major food constituents: the PAEs are extracted into a 1:1 mix of acetonitrile/dichloromethane and glacial acetic acid. After filtration the sample is frozen (-20 °C) to separate the fat. The frozen sample is centrifuged, the supernatant is evaporated to dryness and the residue reconstitute in acetonitrile: 5 μL are analyzed by LC-MS/MS. This procedure will be used also for meat, cheese, orange juices and matrices at high carbohydrate content. This evidences the good variability of such analytical procedure for analyzing PAEs in matrices at different composition. The authors investigated the analytical parameters in the concentration range 0-100 ng $g^{-1}$: the correlation coefficients are above 0.99, the LODs range between 1.7-6.1 ng $g^{-1}$ and the recoveries between 85-115%, the Relative Standard Deviation (RSD) is < 20%.

Finally, in 2014 over the paper by Cao et al. (2014), just reported above in this same section, it should be underlined another very important and complete study by Gu et al. (2014): a sensitive and reliable multi-residue method for determining trace amounts of endocrine disrupting chemicals including five PAEs, five monoalkyphthalate esters (MPEs), four alkylphenols (APs) and bisphenol A (BPA) in seafood is reported. The analytical protocol is based on ultrasonic liquid extraction with acetonitrile followed by solid phase extraction (SPE) with primary secondary amine cartridge for achieving relatively low matrix effects. To authors' knowledge, this is the first study reporting about simultaneous extraction and purification of PAEs, MPEs, APs and BPA in biota samples. For obtaining maximum sensitivity, both LC-MS/MS and GC-MS/MS are applied for the analysis. This interesting method has been validated and tested on fish, mollusk and prawn; the correlation coefficients are > 0.99 for all analytes and good recoveries are obtained, from 60% to

127%. The sensitivity is good with MDLs of 0.015-2.2 ng $g^{-1}$ wet weight (ww) for all compounds.

## Vegetable Matrix

Vegetable is another matrix deeply investigated: almost 50 papers could be found in literature since 2010 (Table 2). This is not a surprise considering that vegetables play an important role in the human diet. For instance, a survey performed few years ago in China revealed some interesting information. The authors (Yang et al. 2011) submitted the population to questionnaires, commonly used to estimate the daily consumption of vegetables by residents. The mean value and standard deviation of ingested rate of each vegetable were calculated according to the questionnaire data. The aggregate ingested rates of total 17 vegetables by males and females were 282.92 and 278.80 g $person^{-1}$ $d^{-1}$, respectively, which were broadly similar to the result of 276.20 g $person^{-1}$ $d^{-1}$ in an investigation by Ministry of Health of the People's Republic of China in 2002 (MHPRC 2010) and 274 g $person^{-1}$ $d^{-1}$ reported by Zhuang et al. (2009). In particular, males consume more leafy vegetables than females in Huizhou but females consume more gourd and fruit vegetables than males. Also, both the amounts of leafy vegetables consumed by males and females were higher than those of gourd and fruit vegetables. On the basis of such data, Shen et al. (2013) determined 16 Polycyclic Aromatic Hydrocarbons (PAHs), 16 PAEs, eight Organochlorine Pesticides (OCPs) and seven Polychlorinated Biphenyls (PCBs) in the 17 frequently-consumed varieties of vegetables. The main extraction occurs by SPE using a C18 cartridge and acetone:dichloromethane (7:3, v:v) as eluent whereas the analysis is performed by GC-MS. Unfortunately, any analytical information are reported except the calibration carried out by seven external standards over the concentration ranges 1-1000 ng $mL^{-1}$ for PAHs, 1-100 μg $mL^{-1}$ for PAEs, 1-100 ng $mL^{-1}$ for OCPs and 1-30 ng $mL^{-1}$ for PCBs.

A very recent paper has been published this year reporting the determination of six priority control PAEs, namely DMP, DEP, DnBP, BBP, DEHP and DnOP, investigated in 44 plastic film greenhouse vegetables and corresponding soils (Wang et al. 2015). Each sample is subjected to cleanup (Ma et al. 2013; Wang et al. 2013): samples are loaded onto a combined column of neutral silica gel and alumina and anhydrous sodium sulfate. Further, the analysis is carried out by means of GC-MS. Under optimum conditions, the recovery rates of spiked soil and vegetable matrix of the six target PAEs range from 76.4 to 109.5% (RSD < 9.1%) in vegetables and soils. The method detection limits of the six target PAEs are 0.03, 0.04, 0.05, 0.02, 0.08 and 0.11 μg $g^{-1}$ for DMP, DEP, DnBP, BBP, DEHP and DnOP, respectively, and the instrumental detection limits range from 0.10 to 0.31 μg $g^{-1}$. By this analytical procedure, the authors determine total PAEs ranging from 0.51 to 7.16 mg $kg^{-1}$ in vegetables and 0.40 to 6.20 mg $kg^{-1}$ in soils with average concentrations of 2.56 and 2.23 mg $kg^{-1}$, respectively. DnBP, DEHP and DnOP contributed more than 90% of the total PAEs in both vegetables and soils but the proportions of DnBP and DnOP in vegetables were significantly higher than in soils.

Of course, many other papers regard this determination but, actually, we have just discussed almost of them in the sections above (Colacino et al. 2010; Sioen et al. 2012; Fierense et al. 2012; Bradley et al. 2013b; Liu et al. 2013; Purcaro et al. 2013; Bradley et al. 2014; Ji et al. 2014; Serrano et al. 2014; Mervish et al. 2014).

## Cheese and Egg Matrices

Few papers regard the PAEs determination in such matrices (Table 2). In particular, Bradley et al. (2013a and 2013b) developed a method for analyzing simultaneously cheese and egg matrices: this method has just described in the above section. We would like to remember that it is based on extraction with acetonitrile/dichloromethane (1:1) and glacial acetic acid and analysis by LC-MS/MS. The analytical parameters have been studied in the concentration range 0-100 ng $g^{-1}$: the correlation coefficients are above 0.99, the LODs range between 1.7-6.1 ng $g^{-1}$ and the recoveries between 85-115%, the Relative Standard Deviation (RSD) is < 20%.

An interesting paper was prepared by Fierens et al. (2013): they determined the levels of eight phthalates (DMP, DEP, DiBP, DnBP, BzBP, DEHP, DCHP, DnOP) in several Belgian milk (milk and milk powder) and its relative dairy products (cheese and butter) observing a PAE contamination at different stages in the milk chain. Actually, the detailed procedure is reported in other paper (Fierens et al. 2012b) but here they show its large potentiality in the analyses of different matrices. The dairy product sample (and milk as well) is extracted with acetone/n-hexane (1:1) to obtain at least 0.5 g fat; subsequently, a Gel Permeation Chromatography (GPC) is applied to purify all fat extracts. The PAE analysis is carried out by means of GC-LR-MS with Electron Impact Ionisation (GC-EI/MS). It should be evidence that they use the same protocol for analyzing the packed and unpacked products. According such procedure, the recoveries vary between 93-100% for high-fat foods and between 82-99% for packaging materials. RSDs are less than 14% for each PAE and the LOQs for milk and dairy products vary between 5-60 μg $kg^{-1}$ fat. This entire analytical procedure was after applied by Van Holderbeke et al. (2014) for analyzing 591 foods and 30 packaging materials sold on the Belgian market. It should be underlined that this paper is one of the most complete investigation of PAE content in different food matrices: fruits and vegetables, milk and dairy products, cereals and cereal products, meat and meat products, fish and fish products, fat and oils, snacks, condiments and sauces, baby food, beverages, vegetarian food, eggs, boiling water (pasta/rice), packaging materials. It is relevant the PAE determination in bread because it is not so common such evaluation and Fierens et al. (2012b) found relatively high DEHP concentration in bread samples (levels up to 1073 μg $g^{-1}$): this level is much higher if we consider the large share of bread in the food consumption pattern of children and adults.

In another paper, just cited for the fish matrix, Cao et al. (2014) uses acetonitrile and a Solid Phase Extraction (SPE) based on a column packed with $MgSO_4$ and Primary Secondary Amine (PSA): after centrifugation, the acetonitrile extract is filtered and concentrated to obtain the final solution analyzed by GC-MS. Using this procedure the authors determine good analytical parameters: the correlation coefficients for eight PAEs (DEP, DiBP, DBP, BBzP, DEHA, DEHP, DOP) are around 0.999 in the concentration range of 0.02-10 ng $mL^{-1}$ with a MDL ranging from 0.0016 μg $g^{-1}$ (DHP) to 0.43 (DEHP) μg $g^{-1}$ whereas the average recoveries for all analytes range between 92-108%.

One of the most interesting and complete paper regarding the PAE evaluation is the study carried out by Ji et al. (2014) where a total of 448 samples including foodstuffs (rice, steamed bun, vegetables, meat, poultry, fish, milk and fruits), ambient PM10, drinking water, soil, indoor PM10 and indoor dust samples from Tianjin were obtained both to determine the distribution of six priority PAEs (DMP, DEP, DBP, BBP, DEHP, DOP) and assess the human exposure to them. The extraction of the food samples are performed by means of a ultrasonic homogenizer instrument and, after concentration under a gentle nitrogen stream to obtain a final volume, the extract is analyzed by GC-MS (in Selected Ion Monitoring, SIM, mode) (Kong et al. 2012). For food samples, the matrix-spike recoveries of PAEs range from 69% to 95% for meat, from 77% to 94% for vegetables and from 87% to 95% for milk; the method precision is good with RSDs between 1.9 and 23.6% for all PAEs. The linearity concentration range has been studied from 1 to 50 ng $mL^{-1}$ for PAEs, the $r^2$ is ≥0.990 and the LOQs for the PAEs considered are between 0.003-0.018 ng $g^{-1}$ fresh weight.

## Fruit Matrix

About 30 articles regarding the PAE migration from packaging and relative determination in fruit, or the association with human health in population according to the International dietary, have been published since 2010 (Table 2). Among these few papers are dedicated to the investigation of the analytical parameters for such matrix. In some cases the PAE content is determined in urine (Toshima et al. 2012; Trasande et al. 2013; Mervish et al. 2014; Bai et al. 2015) or, more generally, in population exposure (Sioen et al. 2012; Bradley et al. 2013a; Serrano et al. 2014): a study found an association between urinary PAE metabolites and increased waist circumference and insulin resistance (Stahlhut et al. 2007; Colacino et al. 2010).

About the analytical protocols, Ma et al. (2010) developed a simple, rapid and sensitive method for the determination of five phthalates (dimethyl phthalate (DMP), diethyl phthalate (DEP), dipropyl phthalate (DPP), benzyl butyl phthalate (BBP), and dicyclohexyl phthalate (DcHP) in fruit jellies based on SPE, termed quick, easy, cheap, effective, rugged, and safe (QuEChERS), and analyzed by LC-MS. The correlation coefficients range from 0.9976 to 0.9991 in the concentration range of 10-250 ng $mL^{-1}$. The LODs and LOQs are in the ranges of 0.09-3.68 ng $mL^{-1}$ and 0.28-11.25 ng $mL^{-1}$, respectively, whereas the recoveries of all five phthalates are in the range of 83.5-103.9% and the RSD values range from 2.0-7.6%.

In 2012, for the first time Zhang et al. (2012) investigated the content of seventeen compounds, including DBP and diisooctyl phthalate (DiOP), in the hydrodistilled oil from the fruit residue of *Ficus Altissima*, which is a species native to India to south China and Malaysia and extensively used as a traditional folk medicinal plant for the treatment of various ailments (i.e., coronary heart disease, senile dementia, cerebral thrombosis, nervous system diseases). Actually, the authors were only interested in the determination of the PAE content (DBP 18.14% and DiOP 7.86% of the oil, respectively) using GC-MS as analytical technique whereas no relevant information are reported for the analytical parameters.

**Table 2. Scheme showing the main protocols used in the PAE determination in different food and beverage matrices. Please, note that papers reporting the determination of the PAE levels with very poor analytical information are cited in the text but not included in this table as well as some papers dealing PAE determination in different matrices using the same protocol**

| Matrix | Extraction | Analytical method | Reference |
|---|---|---|---|
| *Pasta and rice* | SPE with $C_6H_{14}$ | GC×GC with dual FID/MS | Purcaro et al. 2013 |
| *Meat and Fish* | liquid extraction with $C_6H_{14}$ + SPE with Oasis MAX cart. | GC-MS | Guo et al. 2010 |
| | extraction with acetonitrile | GC-MS | Cao et al. 2014 |
| | vortexing after addition of $CH_3CN$ and $MgSO_4$ | LC-MS/MS | Xu et al. 2014 |
| | extraction with $CH_2Cl_2$ | GC-MS | Fierens et al. 2012a, 2012b |
| | ultrasonic extraction with and centrifugation | GC-MS | He et al. 2015 |
| | GPC on Bio beads | LC-UV/MS | Jarošová et al. 2010 |
| | alkaline hydrolysis to PA + derivatization to DMP | GC-FID | Ostrovský et al. 2011 |
| | extraction with $CH_2Cl_2$ | GC-FID | Adeniyi et al. 2011 |
| | extraction with mix of $CH_3CN/CH_2Cl_2$ (1:1) | LC-MS/MS | Bradley et al. 2013a, 2013b |
| | ultrasonic liquid extraction with $CH_3CN$ + SPE with PSA | LC-MS/MS and GC-MS/MS | Gu et al. 2014 |
| *Vegetable* | SPE C18-cart and acetone:$CH_2Cl_2$ | GC-MS | Shen et al. 2013 |
| | clean-up onto silica gel/alumina/$Na_2SO_4$ column | GC-MS | Wang et al. 2015 |
| *Cheese and Egg* | extraction with $CH_3CN/CH_2Cl_2$ + glacial $CH_3COOH$ | LC-MS/MS | Bradley et al. 2013a, 2013b |
| | extraction with acetone/$C_6H_{14}$ + GPC | GC-EI/MS | Fierens et al. 2013 |
| | SPE onto $MgSO_4$ and PSA column | GC-MS | Cao et al. 2014 |
| | ultrasonic homogenization | GC-MS | Ji et al. 2014 |

**Table 2. (Continued)**

| Matrix | Extraction | Analytical method | Reference |
|---|---|---|---|
| *Fruit* | SPE | LC-MS | Ma et al. 2010 |
| | | GC-MS | Zhang et al, 2012 |
| | liquid-liquid extraction + GPC | GC-LR-EI/MS | De Vriese et al. 2005 |
| | extraction with $CH_2Cl_2$ | GC-MS | Bradley et al. 2014 |
| | homogenization with $CH_3OH$, sonication and freezing | LC-MS/MS | Wu et al. 2014 |
| | ultrasonic homogenization | GC-MS | Ji et al. 2014 |
| | Soxhlet using 100% ethanol | GC-MS (NMR, FTIR, UV-Vis) | Anantharaman et al. 2015 |
| *Coffee and Tea* | vortex solvent bar micro-extraction | GC-MS | Huang et al. 2012 |
| | | ESI-HPIMS and ESI-HPIMS | Midey et al. 2013 |
| | SPE using Oasis HLB glass cartridges | GC-MS | Di Bella et al. 2014 |
| | microextraction packed sorbent | ESI-HPIMS | Midey et al. (2013 |
| *Cigarette* | ultrasonic extraction with ethanol | LC-MS/MS | Li et al. 2013 |
| | SDE extraction | GC-MS | Xie et al. 2013 |
| *Beverage* | DLLE | GC-MS | Russo et al. 2012 |
| | DLLE | GC-MS | Russo et al. 2013 |
| | DLLE | GC-MS | Russo et al. 2014a |
| | DLLE | GC-MS | Russo et al. 2014b |
| | DLLME | CE | Sun et al. 2014 |
| | SPE | GC-MS | Al-Saleh and Elkhatib 2014 |
| | Ordered Microporous Carbon | | Lee et al. 2006 |
| | Ordered Microporous Carbon | | Li and Zhao 2013 |
| | MSPE | | Niu et al. 2010 |
| | MSPE | GC-MS | Zhang et al. 2010° |
| | MSPE | HPLC-UV | Wu et al. 2012 |

Numerous studies have indicated that for phthalates, and especially for DEHP, the intake of contaminated foods is the most important exposure pathway for the general population (Wittassek et al. 2011; Wormuth et al. 2006). Due to their possible harmful effects and the importance of dietary intake as exposure route, tolerable daily intakes (TDI) have been specified by the European Food Safety Authority (EFSA) for several phthalates, namely 10 μg $kg^{-1}$ bw $day^{-1}$ for DnBP (EFSA 2005a), 50 μg $kg^{-1}$ bw $day^{-1}$ for DEHP (EFSA 2005b), 500 μg $kg^{-1}$ bw $day^{-1}$ for BBP (EFSA 2005c), and a group TDI of 150 μg $kg^{-1}$ bw $day^{-1}$ for DiNP and DiDP (EFSA 2005d,e). Risk characterization for these phthalates can be conducted by evaluating the assessed dietary intake against these TDI values. Sioen et al. (2012) published the results of a second intensive campaign performed to obtain accurate and sensitive data of eight phthalates (BBP, DnBP, DCHP, DEHP, DEP, DiBP, DMP, DnOP) in all kinds of food products and packaging materials sold on the Belgian market: they analyzed more than 500 food products chosen according to the consumption data from the Belgian national food consumption survey (De Vriese et al. 2005). The authors analyzed a lot of different food products different for fat content: so they had the need to get a methodology to be applied to various cases. The analytical method, just described for the cheese and egg matrices, is based on a liquid-liquid extraction, a purification via gel permeation chromatography (GPC): the PAEs of interest are analyzed by means of GC-LR-EI/MS (Fierens et al. 2012b, 2012c). The LOQs vary between 5 and 145 ng $g^{-1}$ fat for high-fat food samples, between 0.2 and 8.0 ng $g^{-1}$ product for low-fat food samples and between 0.01 and 0.03 ng $g^{-1}$ product for aqueous-based beverages; on the other hand no relevant information are reported about other analytical parameters. Finally, another important problem dealt by authors regards the PAE contamination during the analysis: each analytical sequence is composed of two procedural blanks, several solvent blanks, calibration standards, a reference sample and a limited amount of samples (twelve samples at the most) to reduce the risk of contamination (Fierens et al. 2012b, 2012c).

In 2013, Bradley et al. (2013a) set up a method just extensively described in the cheese and egg matrices: this same method has been successfully applied to the fruit matrix with the same analytical parameters. On the other hand, in another paper Bradley et al. (2014) used a method based on extraction with dichloromethane, centrifugation and analysis by GC-MS for determining DIPN and DiBP in banana and apple fruits (actually, also matrices like mushrooms and potatoes were undergone to the same procedure). Even if the paper is interesting because it regards a study of migration from paper and board into fruit and vegetables and into Tenax™ as a food stimulant, it is really difficult to evaluate the analytical performance: in fact, although the authors cite how they measured recoveries and LOD, they did not report any data about them.

Between April and July 2011, a major incident of phthalate-contaminated foodstuffs occurred in Taiwan. Five major food categories, including sports drinks, fruit beverages, tea drinks, fruit, jam or jelly, and health foods or supplements in tablet or powder form, were contaminated mainly with di-(2-ethylhexyl)phthalate (DEHP) and/or di-isononyl phthalate (DINP). Few studies described the course of this incident: Wu et al. (2014) determined six PAEs, including DEHP, DINP, DIDP, DNOP, DBP, and BBP, in 2731 suspected phthalate-tainted foodstuffs. They considered a concentration of $\geq 1$ ppm for any of the six phthalate chemicals in the foods studied to be defined as positive. The sample is homogenized with methanol, sonicated and then cooled to room temperature; the upper solution is centrifuged at 3500 rpm for 10 minutes and the supernatant analyzed by LC-MS/MS.

Once again, it should be underlined the study carried out by Ji et al. (2014) where among 448 the authors analyzed some fruit samples for determining the PAEs (DMP, DEP, DBP, BBP, DEHP, DOP) content and evaluating the relative human exposure. They performed the PAE extraction by means of a ultrasonic homogenizer instrument and the analysis by GC-MS. The PAE recoveries range from 75% to 97% for fruit; the method precision for all PAEs is good and RSDs range between 1.9-23.6% and LOQs between 0.003-0.018 ng $g^{-1}$ fresh weight.

Finally, it is to be cited the determination carried out by Anantharaman et al. (2015): for the first time they report the presence of dibutyl phthalate in the fruits of *Rubia cordifolia* (Rubiaceae), commonly known as "Indian madder" and used in the ayurvedic medicinal system for the treatment of skin disease and cancer. In addition, the crude extracts have shown to possess antibacterial, antiviral, antineoplastic, and hepatoprotective activities (Kaur et al. 2010). Chemical and structural characterization was carried out using Ultraviolet-Visible spectrophotometer (UV-Vis), Fourier transform infrared (FTIR), GC-MS, Nuclear Magnetic Resonance (NMR). Fresh fruits are subjected to exhaustive dye extraction in a Soxhlet apparatus using 100% ethanol as solvent. The extract is then evaporated up to obtaining dryness; the dried crude extract is subjected to column chromatography (silica gel, 60-120 mesh). After elution with C6H14:Ethyl acetate (4:1), a GC-MS analysis is carried out for determining DBP.

## Coffee and Tea Matrices

Coffee and tea are the most frequently consumed beverages worldwide. Both of them are commercialized in different way: powder simply or powder hermetically sealed between two thin layers of paper coated with a heat seal layer (pods) or packed in a plastic or aluminum cylindrical package (capsules). These last two packaging greatly contributes to the large diffusion of these beverages but simultaneously some problems arise about the PAE migration from the package to the powder. Three interesting methodological papers should be considered for such matrices (Table 2): Huang et al. (2012) used the vortex solvent bar micro-extraction and GC-MS (VSBME-GC-MS) for investigating the PAE content in ice red tea whereas Midey et al. (2013) reached low nanogram levels of PAEs in soy-based bubble tea using high-performance ion mobility spectrometry with electrospray ionization (ESI-HPIMS) and Di Bella et al. (2014) carried out an approach based on SPE using Oasis HLB glass cartridges (they modified an analytical procedure developed by Mohamed and Ammar 2008, for PAE content in black tea, instant coffee and bottled waters), relative elution with methanol and final analysis by means of GC-MS for analyzing 27 PAEs (DMA, DEA, DMP, DEP, DiBA, DBA, DPrP, BB, MGA, DiBP, DBP, MGP, DPP, BPA, BBP, DEHA, DiHepP, BGP, DcHexP, DEHP, DPhP, DOP, DEHT, DEHS, DiNP, DiDP, DNP; this is one of the most complete study in this field) in espresso coffees from capsules, pods and moka pots.

In details, the VSBME-GC-MS method developed Huang et al. (2012) shows very high enrichment factors (over1500) and good reproducibility (< 7.1%) in a rapid extraction within 5 min. Further, it also shows good correlation coefficients ($R^2 > 0.995$) in the linearity range of 0.5-10 $\mu g\ L^{-1}$ as well as good LODs ranging between 0.4 (DBP) and 76 (DMP) ng $L^{-1}$.

On the other side, the Midey et al. (2013) contribution is interesting and original. They combine a Directspray ESI source with HPIMS giving advantages in faster, simpler sample preparation and analysis: in fact, ESI-HPIMS can be exploited for more rapid, sensitive detection of PAE contamination of beverages, which would be especially useful for on-site sample analysis when used with direct ESI. In this way LODs and LOQs are estimated to be sub-ng µL-1 levels (0.017-0.025 ng $\mu L^{-1}$ and 0.056-0.083 ng $\mu L^{-1}$, respectively) without extensive sample handling or preparation. Further, the authors suggest the combination of the Directspray ESI source with other simplified sample preparation methods such as QuEChERS extraction (Anastassiades et al. 2003) or microextraction packed sorbent (MEPS) (Abdel-

Rehim et al. 2004; Chaves et al. 2010) clean-up that remove interfering matrix compounds that could lead to charge suppression (Kebarle and Ho, 1997; Schug and McNair 2002) or mobility coincidences (false positives).

Finally, the ESI-HPIMS approach by Di Bella et al. (2014) for coffee matrix is valid in a large linearity range, 0.010-5.584 μg $mL^{-1}$ with $R^2$ above 0.992 and a RSD below 9.7%, showing very interesting LODs and LOQs ranging between 0.006-3.200 μg $mL^{-1}$ and 0.018-9.100 μg $mL^{-1}$, respectively, depending on the PAE.

## Cigarette Matrix

Two papers investigate the PAE presence in the tipping paper (Li et al. 2013) and in the tea stem added into cigarettes as a new natural tobacco flavor (Xie et al. 2013) (Table 2). In the first paper, a method based on ultrasonic extraction with ethanol followed by LC-MS/MS has been developed for determining 18 phthalate compounds in tipping paper. The method is shown reliable in the linear range of 1-500 ng $mL^{-1}$ with a LOD of 0.01-1.85 μg $g^{-1}$ and recoveries ranging between 82.8-111.6% and RSD less than 10%. The second paper just shows the volatile components characterization of a new natural tobacco flavor obtained from Tieguangyin tea stem: the analytical procedure is based on the SDE extraction and GC/MS analysis. In this way, among 58 volatile compounds identified (i.e., (E)-nerolidol, (E)-geraniol, linalool, linalool oxide, phenylethyl alcohol, dehydrolinalool, terpineol, phytone, phytol, isophytol, n-hexadecanoic acid, ethyl hexadecanoate, ethyl linoleate, ethyl linolenate) DBP has been identified and evaluated. So, the concrete of Tieguangyin tea stem has been evaluated added into cigarettes as a flavoring, enriching the tobacco aroma, reducing the irritancy and improving the aftertaste of cigarette obviously. Unfortunately, no analytical parameters were investigated in the paper.

Finally, a recent study by Wang et al. (2010) indentifies the source emission characteristics for typical semi-volatile organic compounds (SVOCs) observed indoors, evaluates the gas-phase and particle-phase DEHP concentrations under the steady-state condition and estimates the exposure of PAEs. We would like to remember that, as indoor organic compounds, SVOCs are classified organic compounds having a boiling point in the range of 240-400 °C (WHO, 1989). 25% of PAEs used worldwide are consumed in China (McDonald 2002): in particular, the main kinds of plasticizers used in China are DEHP and DBP. Moreover, other 50 kinds of SVOCs, such as di-isodecyl phthalate (DIDP), di-isononyl phthalate (DINP), p-benzoic acid esters, chlorinated paraffins, alkyl sulfonic acid esters, aliphatic binary acid eaters, epoxy esters, trimellitates, phosphates, and so on are produced, and the annual production amount of phthalate is the greatest of all plasticizers (Tao and Liang 2008). The results of this paper show that the production and consumption quantities of plasticizers, coal and cigarettes in China are the highest in the world; indoor plasticizers are higher in China than that of the developed countries which implies that indoor SVOC pollution may be more serious in China. In particular, DEHP remains mainly in the particle-phase under the steady-state condition; the particle-phase concentration of DEHP increases with increasing particle concentration. Actually, the paper does not show any analytical implications in the PAE determination but it is important to cite it for the repercussions in the high human health risk present in China for the large development and abuse of such kinds of compounds in the lifestyle.

## Beverage Matrix

The beverage matrix is very complex: in fact, it ranges from baby beverages (e.g., fruit juices) and milk to soft drinks (e.g., soda, tonic water, cola) and drinks at very different alcoholic grade (e.g., beer, wine, vodka, whisky).

More than 40 papers are present in literature regarding the PAE determination in different beverages (Table 2). Most of the works regarding the PAE determination in such matrix are just cited and/or extensively discussed above (Kappenstein et al. 2012; Fierens et al. 2012b; Sioen et al. 2012; Huang et al. 2012; Bradley et al. 2013a,b; Hsieh et al. 2013; Xu et al. 2014; Wu et al. 2014; Serrano et al. 2014; Sakhi et al. 2014; Chang et al. 2015).

Among the different and traditional methods used for this determination, Russo's group developed a long experience in extraction and determination of toxicological interesting compounds (Russo and Avino 2001; Russo et al. 2002; Russo et al. 2003; Russo et al. 2004; Russo et al. 2007; Russo et al. 2009; Avino et al. 2009; Avino et al. 2011a; Avino et al. 2011b; Russo et al. 2012a; Notardonato et al. 2013; Cinelli et al. 2013; Veschetti et al. 2013; Cinelli et al. 2014a; Russo et al. 2014a; Cinelli et al. 2014b; Notardonato et al. 2014; Russo et al. 2015a; Venditti et al. 2015b). In particular, Russo et al. (2015b) published a review about the extraction and GC-MS determination in foods and, especially in drinks. The same groups studied extensively the application if the SPE extraction with XAD-2 and relative analysis by means of GC-MS or a Dispersive Liquid Liquid Extraction (DLLE) method (Russo et al. 2012b; Cinelli et al. 2013; Russo et al. 2014b; Russo et al. 2014c). The approach reduces the volume of organic solvent required and simultaneously improves the extraction efficiency without the use of a dispersive solvent. Two papers investigates the use of a dispersive solvent in alcoholic matrices (beer, wine, and liquors with a alcohol content ranges between 4 and 40% alcohol by volume) and soft drinks (soda, cola, tonic, sprite, 7up). Use of a vortex is essential for its helping to disperse the extracting solvent. The authors investigate and optimize different parameters, like the extraction solvents, the number of repeated extraction steps and the isolation technique. Using the optimum analytical condition, very good LODs ($\geq$ 0.022 $\mu$g $L^{-1}$) and LOQs ($\geq$ 0.075 $\mu$g $L^{-1}$) are achieved with recoveries ranging from 85% to 100.5% and enrichment factors from 220-to 300-fold. The DLLE approach coupled with sweeping β-cyclodextrin-modified micellar electrokinetic chromatography is also followed by Sun et al. (2014) for analyzing PAEs in white spirit.

Fruit juices are popular beverages regularly consumed by both adults and children. Some evidence suggests that phthalates may leach from PET bottles. Al-Saleh and Elkhatib (2014) have analyzed the presence or assessed the risk of phthalates (DMP, DEP, DBP BBP, DEHP and DOP) in fruit juice (i.e., apple juice) commercially manufactured available in the Saudi market. The evaluation is performed by solid-phase microextraction connected to gas chromatography-mass spectrometry: all PAE levels are above the LOQ of 0.628 $\mu$g $L^{-1}$.

Finally, recently porous carbon materials have received a great deal of attention due to their many applications including gas separation, water purification, catalyst supports, and electrodes for electrochemical double layer capacitors and fuel cells (Lee et al. 2006; Li and Zhao 2013). Ordered Microporous Carbon material (OMC), first synthesized by the Kyotani group (1997), are a relatively new class of carbonaceous materials and a promising and valid candidate as an adsorbent material due to its uniform ordered structure, high pore volume, and high surface area. Magnetic Solid Phase Extraction (MSPE) based on the use of magnetic adsorbents, is a relatively new sample pretreatment technique for PAE analysis (Niu et al. 2010; Zhang et al. 2010a; Wu et al. 2012). Compared with traditional adsorbents, magnetic adsorbents used in MSPE have received considerable attention because of their ease of operation and excellent dispersibility in aqueous solution. The development of magnetic adsorbents with highly efficient extraction capacity has become a hot research field. Up to now, various MSPE adsorbents, such as polymers (Meng et al. 2011; Li et al. 2012), C18 nanoparticles (Zhang et al. 2010a), ionic liquid (Zhang et al. 2010b), activated carbon (Faulconer et al. 2012), carbon nanotubes (Luo et al. 2012) and graphene (Wu et al. 2012) have been explored. Wu et al. (2015) synthesized a magnetic microporous carbon material from sucrose and using a zeolite as a hard template, followed by the in situ chemical co-precipitation of Fe(II) and Fe(III) ions to impart magnetism. The resulting material has been used as sorbent for PAE extraction (DMP, DEP, DAP and DPP); following desorption with acetone, the PAEs were quantified by HPLC equipped with a DAD system. The authors studied the calibration curve in the 0.1-50 ng $mL^{-1}$ concentration range ($R^2 > 0.997$), the RSD (< 6.5%) and the LODs (30 pg $mL^{-1}$ for the 4 PAEs investigated) in the analysis of fruit juice samples: under these optimum conditions they reached recoveries ranging between 90.6% and 102.5%.

## Conclusion

In this review we would like to have reported the main analytical methods based on GC/HPLC analysis useful for determining PAEs in food matrices, both in terms of reproducibility and accuracy. The big issue regarding contamination is relevant, but literature shows different methods to solve it. Furthermore, there are different methods over the GC/HPLC analysis for analyzing PAEs, most of them based on ELISA procedure (Zhang et al. 2013); in any case, the chromatographic methods still remain the main analytical procedure for such investigation.

## References

Abdel-Rehim, M; Altun, Z; Blomberg, L. *J Mass Spectrom*, 2004, 39, 1488-1493.

Adeniyi, AA; Okedeyi, OO; Yusuf, KA. *Environ Monit Assess*, 2011, 172, 561-569.

Al-Saleh, I; Elkhatib, R. *J Food Meas Character*, 2014, 8, 373-380.

Anantharaman, A; Priya, RR; Hemachandran, H; Akella, S; Rajasekaran, C; Ganesh, J; Fulzele, DP; Siva, R. *Environ Toxicol*, 2015, doi: 10.1002/tox.22115

Anastassiades, M; Lehotay, SJ; Stajnbaher, D; Schenck, FJ. *J AOAC Int*, 2003, 86, 412-431.

Armbruster, DA; Tillman, MD; Hubbs, LM. *Clin Chem*, 1994, 40, 1233-1238.

Armbruster, DA; Pry, T. *Clin Biochem Rev*, 2008, 29, S49-S52.

Ash, M; Ash, I. A formulary adhesives and other sealants; Chemical Publishing Co., Inc.: New York, NY, 1987; 1-365. ISBN 0-8206-0297-3

Aurela, B; Kulmala, H; Soderhjelm, L. *Food Addit Contam*, 1999, 16, 571-577.

Avino, P; Notardonato, I; Cinelli, G; Russo, MV. *Curr Anal Chem*, 2009, 5, 339-346.

Avino, P; Cinelli, G; Notardonato, I; Russo, MV. *Aerosol Air Qual Res*, 2011a, 11, 783-790.

Avino, P; Cinelli, G; Notardonato, I; Russo, MV. *Anal Bioanal Chem*, 2011b, 400, 3561-3571.

Aznar, M; Vera, P; Canellas, E; Nerín, C; Mercea, P; Störmer, A. *J Mater Chem*, 2011, 21, 4358-4370.

Bai, PY; Wittert, GA; Taylor, AW; Martin, SA; Milne, RW; Shi, Z. *Plos One*, 2015, 10, e0122140.

Bart, JC.J Polymer additive analytics: industrial practice and case studies; Monografie Scienze Tecnologiche 14; Firenze University Press: Florence, IT, 2006, 67-165.

Bizarro, C; Ros, O; Vallejo, A; Prieto, A; Etxebarria, N; Cajaraville, MP; Ortiz-Zarragoitia, M. *Marine Environ Res*, 2014, 96, 19-28.

Boccacci Mariani, M; Chiacchierini, E; Gesumundo, C. *Food Addit Contam*, 1999, 16, 207-213.

Boqué, R; Heyden, YV. LC-GC Eur 2009, 22 (available at http://www.chromatographyonline.com/limit-detection; last access on June 2015).

Bosnir, J; Puntarić, D; Skes, I; Klarić, M; Simić, S; Zorić, I. *Coll Antropol*, 2003, 27, 23-30.

Bouma, K; Schakel, DJ. *Food Addit Contam*, 2002, 19, 602-610.

Bradley, EL; Leon, I; Mortimer, D; Speck, DR; Castle, L. *Food Addit Contam*, A 2013a, 30, 722-734.

Bradley, EL; Burden, RA; Bentayeb, K; Driffield, M; Harmer, N; Mortimer, DN; Speck, DR; Ticha, J; Castle, L. *Food Addit Contam A*, 2013b, 30, 735-742.

Bradley, EL; Castle, L; Speck, DR. *Food Addit Contam A*, 2014, 31, 1301-1309.

Cao, X.-L. *Compr Rev Food Sci F*, 2010, 9, 21-43.

Cao, X.-L; Zhao, W; Churchill, R; Hilts, C. *J Food Protect*, 2014, 77, 610-620.

Carnevali, O; Tosti, L; Speciale, C; Peng, C; Zhu, Y; Maradonna, F. *Plos One*, 2010, 5, e10201.

Chang, L; Bi, P; Li, X; Wei, Y. *Food Chem*, 2015, 177, 127-133.

Chaves, AR; Leandro, FZ; Carris, JA; Queiroz, ME.C. *J Chromatogr B*, 2010, 878, 2123-2129.

Chen, P; Li, S; Liu, L; Xu, N. *Environ Toxicol Chem*, 2015, 34, 518-526.

Christensson, A; Ljunggren, L; Nilsson-Thorell, C; Arge, B; Diehl, U; Hagstam, KE; Lundberg, M. *Int J Artif Organs*, 1991, 14, 407-410.

Cinelli, G; Avino, P; Notardonato, I; Centola, A; Russo, MV. *Anal Chim Acta*, 2013, 769, 72-78.

Cinelli, G; Avino, P; Notardonato, I; Centola, A; Russo, MV. *Food Chem*, 2014a, 146, 181-187.

Cinelli, G; Avino, P; Notardonato, I; Russo, MV. *Anal Meth*, 2014b, 6, 782-790.

Clara, M; Windhofer, G; Hartl, W; Braun, K; Simon, M; Gans, O; Scheffknecht, C; Chovaneca, A. *Chemosphere*, 2010, 78, 1078-1084.

Colacino, JA; Harris, TR; Schecter, A. *Environ Health Perspect*, 2010, 118, 998-1003.

Crossfire Beistein Database Elsevier Information Systems GmbH; Frankfurt, Germany. 2009. Available at http://www.info.crossfirebeilstein.com/ (last access on June 2015).

De Vriese, S; De Backer, G; De Henauw, S; Huybrechts, I; Kornitzer, R; Leveque, A; Moreau, M; Van Oyen, H. *Arch Publ Health*, 2005, 63, 1-16.

Di Bella, G; Potortì, AG; Lo Turco, V; Saitta, M; Dugo, G. *Food Control*, 2014, 41, 185-192.

Dolan, JW. *LC-GC Eur*, 2006, 19, 12-16.

EFSA. Opinion of the Scientific Panel on food additives, flavourings, processing aids and material in contact with food (AFC) on a request from the Commission related to di-butylphthalate (DBP) for use in food contact materials: question EFSA-Q-2003-192. *EFSA J*, 2005, 242, 1-2.

EFSA. Opinion of the Scientific Panel on food additives, flavourings, processing aids and materials in contact with food (AFC) on a request from the Commission related to bis(2-ethylhexyl)phthalate (DEHP) for use in food contact materials: question EFSA-Q-2003-191. *EFSA J*, 2005b, 243, 1-2.

EFSA. Opinion of the Scientific Panel on food additives, flavourings, processing aids and materials in contact with food (AFC) on a request from the Commission related to butylbenzylphthalate (BBP) for use in food contact materials: question nEFSA-Q-2003-190. *EFSA J*, 2005c, 241, 1-2.

EFSA. Opinion of the Scientific Panel on food additives, flavourings, processing aids and material in contact with food (AFC) on a request from the Commission related to di-isodecylphthalate (DiDP) for use in food contact materials: question EFSA-Q-2003-195. *EFSA J*, 2005d, 245, 1-2.

EFSA. Opinion of the Scientific Panel on food additives, flavourings, processing aids and material in contact with food (AFC) on a request from the Commission related to di-isononylphthalate (DiNP) for use in food contact materials: question EFSA-Q-2003-194. *EFSA J*, 2005e, 244, 1-2.

Fatoki, OS; Bornman, M; Ravandhalala, L; Chimuka, L; Genthe, B; Adeniyi, A. *Water SA*, 2010, 36, 117-125.

Faulconer, EK; von Reitzenstein, NV; Mazyck, DW. *J Hazard Mater*, 2012, 199, 9-14.

Fierens, T; Vanermen, G; Van Holderbeke, M; De Henauw, S; Sioen, I. *Food Chem Toxicol*, 2012a, 50, 4428-4435.

Fierens, T; Servaes, K; Van Holderbeke, M; Geerts, L; De Henauw, S; Sioen, I; Vanermen, G. *Food Chem Toxicol*, 2012b, 50, 2575-2583.

Fierens, T; Van Holderbeke, M; Willems, H; De Henauw, S; Sioen, I. *Food Chem Toxicol*, 2012c, 50, 2945-2953.

Fierens, T; Van Holderbeke, M; Willems, H; De Henauw, S; Sioen, I. *Environ Int*, 2013, 51, 1-7.

Fierens, T; Standaert, A; Cornelis, C; Sioen, I; De Henauw, S; Willems, H; Bellemans, M; De Maeyer, M; Van Holderbeke, M. *Environ Int*, 2014, 73, 117-127.

Fromme, H; Küchler, T; Otto, T; Pilz, K; Müller, J; Wenzel, A. *Water Res*, 2002, 36, 1429-1438.

Giust, JA. *J Agric Food Chem*, 1990, 38, 415-418.

Gu, Y.-Y; Yu, X.-J; Peng, J.-F; Chen, S.-B; Zhong Y.-Y; Yin, D.-Q; Hu, X.-L. *J Chromatogr B*, 2014, 965, 164-172.

Guo, Z; Wang, S; Wei, D; Wang, M; Zhang, H; Gai, P; Duan, J. *Meat Sci*, 2010, 84, 484-490.

Hansen, OG. *Med Device Technol*, 2007, 18, 10-12.

Hauser, R; Calafat, AM. *Occ Env Med*, 2005, 62, 806-818.

He, M; Yang, C; Geng, R; Zhao, X; Hong, L; Piao, X; Chen, T; Quinto, M; Li, D. *Anal Chim Acta*, 2015, 879, 63-68.

Hillman, LS., Goodwin, SL; Sherman, WR. *N Engl J Med*, 1975, 292, 381-386.

HSDB. Hazardous Substance Data Bank. US. National Library of Medicine TOXNET System. 2009. Available at http://toxnet.nlm.nih.gov/cgi-bin/sis/htmlgen?HSDB (last access on June 2015).

Hsieh, S.-Y; Wang, C.-C; Wu, S.-M. *Food Chem*, 2013, 141, 3486-3491.

Huang, G; Li, H.-F; Zhang, B.-T; Ma, Y; Lin, J.-M. *Talanta*, 2012, 100, 64-70.

Hubingerc JC. *J Cosmet Sci*, 2010, 61, 457-465.

Jarošová, A; Harazim, J; Krátká, L; Kolenčíková, D. *Environ Chem Lett*, 2010, 8, 387-391.

Ji, Y; Wang, F; Zhang, L; Shan, C; Bai, Z; Sun, Z; Liu, L; Shen, B. *J Hazard Mater*, 2014, 279, 133-140.

Kambia, K; Dine, T; Gressier, B; Germe, AF; Luyckx, M; Brunet, C; Michaud, L; Gottrand F. *J Chromatogr B*, 2001, 755, 297-303.

Kappenstein, O; Vieth, B; Luch, A; Pfaff, K. *EXS*, 2012, 101, 87-106.

Karle, VA; Short, BL; Martin, GR; Bulas, DI; Getson, PR; Luban, NL; O'Brien, AM; Rubin, RJ. *Crit Care Med*, 1997, 25, 696-703.

Kaur, P; Chandela, M; Kumar, S; Kumar, N; Singh, B; Kaur, S. *Food Chem Toxicol*, 2010, 48, 320-325.

Kebarle, P; Ho, Y. On the mechanism of electrospray mass spectrometry: fundamentals, instrumentation, and applications. in: RB. Cole (Ed.), Electrospray Ionization Mass Spectrometry; Wiley, New York, NY; 1997, pp. 42-50.

Kim, S; Kang S; Lee G; Lee S; Jo A; Kwak K; Kim D; Koh D; Kho YL; Kim S; Choi, K. *Sci Total Environ*, 2014, 472, 49-55.

Knoll, JK. *J Chromatogr Sci*, 1985, 23, 422-425.

Kong, S; Ji, Y; Liu, L; Chen, L; Zhao, X; Wang, J; Bai, Z; Sun, Z. *Environ Pollut*, 2012, 170, 161-168.

Kyotani, T; Nagai, T; Inoue, S; Tomita, A. *Chem Mater*, 1997, 9, 609-615.

Latini, G. *Clin Chim Acta*, 2005, 361, 20-29.

Latini, G; Del Vecchio, A; Massaro, M; Verrotti, A; De Felice, C. *Toxicol*, 2006, 226, 90-98.

Law, R. (2015) Determination of phthalate esters in soft drinks. Available at http://acceleratingscience.com/polymers2plastics/determination-of-phthalate-esters-in-soft-drinks/ (last access on June 2015).

Lee, J; Kim, J; Hyeon, T. *Adv Mater*, 2006, 18, 2073-2094.

Li, XS; Xu, LD; Shan, YB; Yuan, BF; Feng, YQ. *J Chromatogr A*, 2012, 1265, 24-30.

Li, W; Zhao, DY. *Chem Commun*, 2013, 49, 943-946.

Li, Y; Wang, H; Liao, X; Xiao, S. *Tob Sci Technol*, 2013, 12, 31-35.

Liu, Y., Wang, S., Wang, L. J Agric Food Chem 2013, 13, 1160-1164.

Loff, S; Kabs, F; Witt, K; Sartoris, J; Mandl, B; Niessen, KH; Waag, KL. *J Pediatr Surg*, 2000, 35, 1775-1781.

Luo, YB; Yu, QW; Yuan, BF; Feng, YQ. *Talanta*, 2012, 90, 123-131.

Ma, Y; Hashi, Y; Ji, F; Lin, JM. *J Sep Sci*, 2010, 33, 251-257.

Ma, TT; Christie, P; Luo, YM; Teng, Y. *Environ Geochem Health*, 2013, 35, 465-476.

Mackintosh, CE; Maldonado, JA., Ikonomou, MG; Gobas, FAPC. *Environ Sci Technol*, 2006, 40, 3481-3488.

Manilal, A; Sujith, S; Seghal Kiran, G; Selvin, J; Panikkar, MVN. *Thalassas*, 2011, 27, 47-56.

McDonald, TA. *Chemosphere*, 2002, 46, 745-755.

Meng, J; Bu, J; Deng, C; Zhang, X. *J Chromatogr A*, 2011, 1218, 1585-1591.

Mervish, N; McGovern, KJ; Teitelbaum, SL; Pinney, SM; Windham, GC; Biro, FM; Kushi, LH; Silva, MJ; Ye, X; Calafat, AM; Wolff, MS; BCERP. Dietary predictors of urinary environmental biomarkers in young girls, BCERP, 2004-7. *Environ Res*, 2014, 133, 12-19.

Midey, AJ; Camacho, A; Sampathkumaran, J; Krueger, CA; Osgood, MA; Wu, C. *Anal Chim Acta*, 2013, 804, 197-206.

Miller, JC; Miller, JN. Statistics for analytical chemistry. *Ellis Horword; Chichester*, UK; 1998.

Ministry of Health of the People's Republic of China (MHPRC). Year Book of China Health Statistics 2010. Publishing House of Peking Union Medical College, Beijing, China; 2010.

Mohamed, MA; Ammar, AS. *Am J Food Technol*, 2008, 3, 341-346.

Niu, H; Zhang, S; Zhang, X; Cai, Y. *ACS Appl Mat Interfaces*, 2010, 2, 1157-1163.

Notardonato, I; Avino, P; Centola, A; Cinelli, G; Russo, MV. *Anal Bioanal Chem*, 2013, 405, 6137-6141.

Notardonato, I; Avino, P; Cinelli, G; Russo, MV. *RSC Adv*, 2014, 4, 42424-42431.

Oie, L; Hersoug LG; Madsen, JO. *Environ Health Perspect*, 1997, 105, 972-978.

Ostrovský, I; Čabala, R; Kubinec, R; Górová, R; Blaško, J; Kubincová, J; Řimnáčová, L; Lorenz, W. *Food Chem*, 2011, 124, 392-395.

Petersen, JH; Breindahl, T. *Food Addit Contam*, 2000, 17, 133-141.

Purcaro, G; Tranchida, PQ; Barp, L; Moret, S; Conte, LS; Mondello, L. *Anal Chim Acta*, 2013, 773, 97-104.

Rodgers KM; Rudel, RA; Just, Allan C. Phthalates in food packaging, consumer products, and indoor environments. Toxicants in Food Packaging and Household Plastics; Ed. Snedeker, SM; Springer: London, UK; 2014, 31-59. ISBN: 978-1-4471-6499-9.

Roth, B; Herkenrath, P; Lehmann, HJ.,; Ohles, HD; Hömig, HJ; Benz-Bohm, G; Kreuder, J; Younossi-Hartenstein, A. *Eur J Pediatr*, 1988, 147, 41-46.

Russo, MV; Avino, P. *Anal Lett*, 2001, 34, 1003-1013.

Russo, MV; Campanella, L; Avino, P. *J Chromatogr B*, 2002, 780, 431-441.

Russo, MV; Campanella, L; Avino, P. *J Sep Sci*, 2003, 26, 376-380.

Russo, MV; Di Lorenzo, V; Avino, P. *Ann Chim*, 2004, 94, 741-749.

Russo, MV; Veschetti, E; Cinelli, G; Avino, P. *Chromatographia*, 2007, 66, 237-242.

Russo, MV; Avino, P; Notardonato, I; Cinelli, G. *Chromatographia*, 2009, 69, 709-717.

Russo, MV; Avino, P; Cinelli, G; Notardonato, I. *Anal Bioanal Chem*, 2012a, 404, 1517-1527.

Russo, MV; Notardonato, I; Cinelli, G; Avino, P. *Anal Bioanal Chem*, 2012b, 402, 1373-1381.

Russo, MV; Avino, P; Centola, A; Notardonato, I; Cinelli, G. *Food Chem*, 2014a, 146, 204-211.

Russo, MV; Notardonato, I; Avino, P; Cinelli, G. *RSC Adv*, 2014b, 4, 59655-59663.

Russo, MV; Notardonato, I; Avino, P; Cinelli, G. *Anal Meth*, 2014c, 6, 7030-7037.

Russo, MV; Avino, P; Veschetti, E. *RSC Adv*, 2015a, 5, 10418-10423.

Russo, MV; Avino, P; Perugini, L; Notardonato, I. *RSC Adv*, 2015b, 5, 37023-37043.

Sakhi, AK; Lillegaard, IT.L; Voorspoels, S; Carlsen, MH; Løken, EB; Brantsæter, AL; Haugen, M; Meltzer, HM; Thomsen, C. *Environ Int*, 2014, 73, 259-269.

Sánchez-Avila, J; Tauler, R; Lacorte, S. *Environ Int*, 2012, 46, 50-62.

Sathyanarayana, S; Karr, CJ; Lozano, P; Brown, E; Calafat, AM; Liu, F; Swan, SH. *J Pediatrics*, 2008, 121, e260-e268.

Schettler, T. *Int J Andrology*, 2006, 29, 134-139.

Schug, K; McNair, HM. *J Sep Sci*, 2002, 25, 759-766.

Serrano, SE; Karr, CJ; Seixas, NS; Nguyen, RH; Barrett, ES; Janssen, S; Redmon B; Swan, SH; Sathyanarayana, S. *Int J Environ Res Pub Health*, 2014, 11, 6193-6215.

Sharman, M; Read, WA; Castle, L; Gilbert, J. *Food Addit Contam*, 1994, 11, 375-385.

Shea, KM. *Pediatrics*, 2003, 111, 1467-1474.

Shen, L; Xia, B; Dai, X. *Chemosphere*, 2013, 93, 2254-2263.

Sioen, I; Fierens, T; Van Holderbeke, M; Geerts, L; Bellemans, M; De Maeyer, M; Servaes, K; Vanermen, G; Boon, PE; De Henauw, S. *Environ Int*, 2012, 48, 102-108.

Sjoberg, P; Bondesson, UG; Sedin, EG; Gustafsson, JP. *Eur J Clin Invest*, 1985a, 15, 430-436.

Sjoberg, PO; Bondesson, UG; Sedin, EG; Gustafsson, JP. *Transfusion*, 1985b, 25, 424-428.

Stahlhut, RW; van Wijngaarden, E; Dye, TD; Cook, S; Swan, SH. *Environ Health Perspect*, 2007, 115, 876-882.

Steiner, I; Scharf, L; Fiala, F; Washüttl, J. *Food Addit Contam*, 1998, 15, 812-817.

Stringer, R., Labunska, I; Santillo, D; Johnston, P; Siddorn, J; Stephenson, A. *Environ Sci Pollut Res Int*, 2000, 7, 27-36.

Sturaro, A; Parvoli, G; Rella, R; Bardati, S; Doretti, L. *Int J Food Sci Technol*, 1994, 29, 593-603.

Summerfield, W; Cooper, I. *Food Addit Contam*, 2001, 18, 77-88.

Sun, J; He, H; Liu, S. *J Sep Sci*, 2014, 37, 1679-1686.

Tao, G; Liang, C. *Plastic Sci Tech*, 2008, 36, 78-81.

Teil, MJ; Blanchard, M; Chevreuil, M. *Sci Total Environ*, 2006, 354, 212-223.

Tickner, JA., Schettler, T; Guidotti, T; McCally, M; Rossi, M. *Am J Ind*, 2001, 39, 100-111.

Toshima, H; Suzuki, Y; Imai, K; Yoshinaga, J; Shiraishi, H; Mizumoto, Y; Hatakeyama, S; Onohara, C; Tokuoka, S. *Int J Hyg Environ Health*, 2012, 215, 502-506.

Trasande, L; Sathyanarayana, S; Messito, MJ; Gross, RS; Attina, TM; Mendelsohn AL. *Environ Res*, 2013, 126, 84-90.

Tyrrell, J; Melzer, D; Henley, W; Galloway, TS; Osborne, N. *J. Environ Inter*, 2013, 59, 328-335.

U.S. Agency for Toxic Substances and Disease Registry (US ATSDR). Toxicological profile for diethyl phthalate. 1995. Available at http://www.atsdr.cdc.gov/toxprofiles/tp73.html (last access on June 2015).

U.S. Agency for Toxic Substances and Disease Registry (US ATSDR). Toxicological profile for di-n-Butyl phthalate. 1997a. Available at http://www.atsdr.cdc.gov/toxprofiles/tp135.html (last access on June 2015).

U.S. Agency for Toxic Substances and Disease Registry (US ATSDR). Toxicological profile for di-n-Octylphthalate (DNOP). 1997b. Available at http://www.atsdr.cdc.gov/toxprofiles/tp95.html (last access on June 2015).

U.S. Agency for Toxic Substances and Disease Registry (US ATSDR). Toxicological profile for di(2-ethylhexyl)phthalate (DEHP). 2002. Available at http://www.atsdr.cdc.gov/toxprofiles/tp9.html (last access on June 2015).

U.S. Environmental Protection Agency (US EPA). Ground Water and drinking water. Technical Fact Sheet on: di(2-Ethylhexyl)phthalate (DEHP). 2002. Available at http://www.epa.gov/safewater/dwh/t-soc/dehp.html (last access on June 2015).

U.S. National Toxicology Program (US NTP). Center for the evaluation of risks to human reproduction. NTP-CERHR Reports and Monographs. 2007. Available at http://cerhr.niehs.nih.gov/reports/index.html (last access on June 2015).

Van Vliet, ED; Reitano, EM; Chhabra, JS; Bergen, GP; Whyatt, RM. *J Perinatol*, 2011, 31, 551-560.

Venditti, F; Cuomo, F; Ceglie, A; Avino, P; Russo, MV; Lopez, F. *Langmuir*, 2015b, 31, 3627-3634.

Vera, P; Aznar, M; Mercea, P; Nerín, C. *J Mater Chem*, 2011, 21, 420-431.

Veschetti, E; Notardonato, I; Avino, P; Russo, MV. *J Sep Sci*, 2013, 36, 2260-2267.

Vitali, M; Guidotti, M; Macilenti, G; Cremisini, C. *Environ Int*, 1997, 23, 337-347.

Wang, L; Zhao, B; Liu, C; Lin, H; Yang, X; Zhang, Y. *Chinese Sci Bull*, 2010, 55, 1469-1478.

Wang, J; Luo, YM; Teng, Y; Ma, WT; Christie, P; Li, ZG. *Environ Pollut*, 2013, 180, 265-273.

Wang, J; Chen, G; Christie, P; Zhang, M; Luo, Y; Teng, Y. *Sci Total Environ*, 2015, 523, 129-137.

Wigle, DT; Arbuckle, TE; Turner, MC; Berube, A; Yangm Q; Lium S; Krewskim D. *J Toxicol Environ Health B Crit Rev*, 2008, 11, 373-517.

Wilkinson, CF; Lamb, JC. *Regul Toxicol Pharmacol*, 1999, 30, 140-155.

Wittassek, M; Koch, HM; Angerer, J; Bruning, T. *Mol Nutr Food Res*, 2011, 55, 7-31.

World Health Organization (WHO). Indoor air quality: Organic pollutants. EURO Reports and Studies No III; World Health Organization; Copenhagen, DK; 1989.

Wormuth, M; Scheringer, M; Vollenweider, M; Hungerbuhler, K. Risk Anal 2006, 26, 803-824.

Wu, B; Ford, T; Gu, J.-D; Zhang, XX; Li, A.-M; Cheng, S.-P. *Chemosphere*, 2010, 80, 535-541.

Wu, QH; Liu, M; Ma, XX; Wang, WN; Wang, C; Zang, XH; Wang, Z. *Microchim Acta*, 2012, 177, 23-30.

Wu, C.-F; Chang-Chien, G.-P; Su, S.-W; Chen, B.-H; Wu, M.-T. *J Formos Med Assoc*, 2014, 113, 600-605.

Wu, Q; Zhou, X; Sun, M; Ma, X; Wang, C; Wang, Z. *Microchim Acta*, 2015, 182, 879-885.

Xie, Q; Li, J; Wang, K; Dong, Y; Yao, R; Dong, D. *Tob Sci Technol*, 2013, 7, 48-54.

Xu, D; Deng, X; Fang, E; Zheng, X; Zhou, Y; Lin, L; Chen, L; Wu, M; Huang, Z. *J Chromatogr A*, 2014, 1324, 49- 56.

Yang, QW; Xu, Y; Liu, SJ; He, JF; Long, FY. *Ecotoxicol Environ Saf*, 2011, 74, 1664-1669.

Zhang, S; Niu, H; Cai, Y; Shi, Y. *Anal Chim Acta*, 2010a, 665, 167-175.

Zhang, QL; Yang, F; Tang, F; Zeng, K; Wu, KK; Cai, QY; Yao, SZ. *Analyst*, 2010b, 135, 2426-2433.

Zhang, M; Yu, X; Wang, Y; Hu, Y; Liu, S. *Food Anal Meth*, 2013, 6, 1223-1228.

Zhuang, P; McBride, MB., Xia, H; Li, N; Li, Z. *Sci Total Environ*, 2009, 407, 1551-1561.

In: Phthalates
Editor: Miguel Owens
ISBN: 978-1-63483-908-2


*Chapter 3*

# The Role of Liquid Chromatography-Mass Spectrometry Techniques in the Detection, Determination and Confirmation of Phthalates

***Paolo Lucci***[1] ***and Oscar Núñez***[2,3,*]
[1]Department of Food Science,
University of Udine, Udine, Italy
[2]Department of Analytical Chemistry,
University of Barcelona, Barcelona, Spain
[3]Serra Húnter Fellow, Generalitat de Catalunya, Spain

## Abstract

Phthalates (diesters of 1,2-benzenedicarboxylic acid) are a group of anthropogenic chemicals that are broadly employed in many household and personal care products, toys, and medical supplies, as well as plasticizers in various domestic and industrial products to increase the flexibility, durability, and transparency of plastics such as polyvinyl chloride (PVC) and cellulose acetate. The widespread use and the application of these compounds have resulted in their ubiquitous presence in the environment, and because they are classified by many countries (including EU and USA) as carcinogenic, mutagenic and toxic to

[*] E-mail address: oscar.nunez@ub.edu

reproduction, human exposure to phthalates is currently receiving considerable attention in political and scientific circles. These compounds are known to be endocrine disruptors, man-made chemicals that disrupt the endocrine system by competing with endogenous hormones by binding to receptors or by altering the synthesis and metabolism of these hormones. Epidemiological studies on human exposure to phthalates from different countries have shown that increased phthalate exposure might have associations with potential health risks such as allergic diseases in children such as asthma and eczema, reproductive effects (male fertility), precocious puberty in pubertal girls, increased incidence of chronic disease, and a possible role in the development of cancer.

Since 14 December 2005, the EU Directive 2005/84/EC has banned several phthalates for use in PVC and other plasticized materials in all toys and childcare articles. Moreover, several studies have concluded that diet is the major route of exposure for humans, and that environmental contamination is one of the sources of these chemicals in food at various levels. So, the development of analytical methodologies for the monitoring of these compounds in a great variety of matrices including environmental ones, food products and food packaging materials, as well as in biological fluids for epidemiological studies on human exposure is mandatory.

Various analytical methods have been developed for measuring phthalates, mostly using gas chromatography coupled with mass spectrometry (GC-MS) and liquid chromatography coupled with mass spectrometry (LC-MS) or tandem mass spectrometry (LC-MS/MS). LC-MS(/MS) methods, which combine high selectivity and sensitivity, show an advantage over GC-MS methods because of simple sample preparation without the need for derivatization. Recently, high resolution mass spectrometry (HRMS) is acquiring also an important role in the analysis of phthalates by taking advantage of the higher resolving power and higher accurate mass measurements achieved, and today several LC-HRMS methods are reported for the quantitative determination of these compounds.

In this chapter the role of liquid chromatography-mass spectrometry in the detection, determination and confirmation of phthalates in a variety of matrices (environmental, food and bio-analytical ones) will be reviewed. LC-MS(/MS) chromatographic conditions, ionization sources, and MS and HRMS analyzers frequently used, as well as strategies for the structural characterization and the qualitative and quantitative analysis of phthalates and their metabolites will be discussed by means of relevant applications. Coverage of all kind of applications is beyond the scope of the present contribution, so we will focus on the most relevant applications published in the last years.

## INTRODUCTION

Diesters of 1,2-benzenedicarboxylic acid (phthalic acid), commonly referred to as phthalate esters (PAEs) or simply phthalates, are synthetic, man-made chemicals of increasing public importance because of their potential toxicity [1, 2]. Figure 1 depicts the general chemical structure of phthalates, and Table 1 shows a list of the most common phthalate compounds. PAEs have been widely used for more than 50 years as plasticizers and they are essential for improving plasticity, flexibility, and elasticity [3, 4]. Therefore, these chemicals have a wide spectrum of industrial applications and they also appear in a wide range of consumer products, as well as in food processing and in medical applications [5-7]. Due to its utility and cost-effectiveness, approximately 2.0-4.3 million tons per year worldwide are produced [8]. Phthalate plasticizers are not chemically bound to plastics such as polyvinyl chloride (PVC), so due to their non-covalent bonding properties, high production volume and widespread use, phthalates are released extensively in the environment [9].

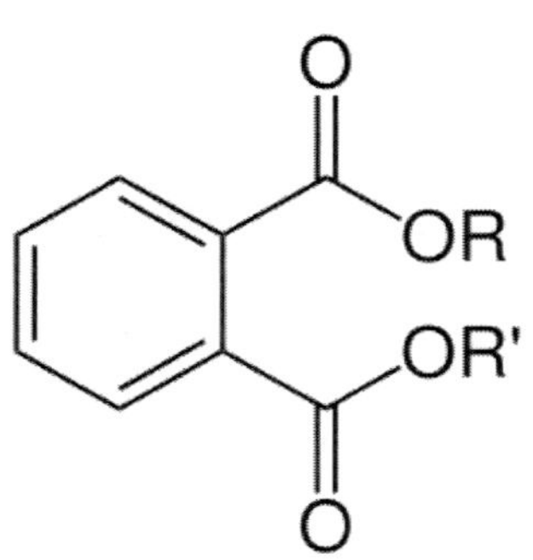


Figure 1. General chemical structures of phthalates (R and R' are general groups).

Because of these reasons, PAEs have been considered as one of the priority pollutants and endocrine disrupting chemicals by the US Environmental Protection Agency and other regulatory agencies in several countries [10, 11].

**Table 1. List of the most common phthalate compounds**

| Name | Abbreviation | Structural formula | CAS No. |
|---|---|---|---|
| Dimethyl phthalate | DMP | $C_6H_4(COOCH_3)_2$ | 131-11-3 |
| Diethyl phthalate | DEP | $C_6H_4(COOC_2H_5)_2$ | 84-66-2 |
| Diallyl phthalate | DAP | $C_6H_4(COOCH_2CH{=}CH_2)_2$ | 131-17-9 |
| Di-n-propyl phthalate | DPP, DPrP | $C_6H_4[COO(CH_2)_2CH_3]_2$ | 131-16-8 |
| Di-n-butyl phthalate | DBP, DnBP | $C_6H_4[COO(CH_2)_3CH_3]_2$ | 84-74-2 |
| Diisobutyl phthalate | DIBP | $C_6H_4[COOCH_2CH(CH_3)_2]_2$ | 84-69-5 |
| Butyl cyclohexyl phthalate | BCP | $CH_3(CH_2)_3OOC\text{-}C_6H_4\text{-}COOC_6H_{11}$ | 84-64-0 |
| Di-n-pentyl phthalate | DNPP, DPeP | $C_6H_4[COO(CH_2)_4CH_3]_2$ | 131-18-0 |
| Dicyclohexyl phthalate | DCP, DcHP | $C_6H_4[COOC_6H_{11}]_2$ | 84-61-7 |
| Benzyl butyl phthalate | BBP | $CH_3(CH_2)_3OOC\text{-}C_6H_4\text{-}COOCH_2C_6H_5$ | 85-68-7 |
| Di-n-hexyl phthalate | DNHP, DHxP | $C_6H_4[COO(CH_2)_5CH_3]_2$ | 84-75-3 |
| Diisohexyl phthalate | DIHxP | $C_6H_4[COO(CH_2)_3CH(CH_3)_2]_2$ | 146-50-9 |
| Diisoheptyl phthalate | DIHpP | $C_6H_4[COO(CH_2)_4CH(CH_3)_2]_2$ | 41451-28-9 |
| Butyl decyl phthalate | BDP | $CH_3(CH_2)_3OOC\text{-}C_6H_4\text{-}COO(CH_2)_9CH_3$ | 89-19-0 |
| Di(2-ethylhexyl) phthalate | DEHP, DOP | $C_6H_4[COOCH_2CH(C_2H_5)(CH_2)_3CH_3]_2$ | 117-81-7 |
| Di(n-octyl) phthalate | DNOP, DnOP | $C_6H_4[COO(CH_2)_7CH_3]_2$ | 117-84-0 |
| Diisooctyl phthalate | DIOP | $C_6H_4[COO(CH_2)_5CH(CH_3)_2]_2$ | 27554-26-3 |
| n-Octyl n-decyl phthalate | ODP | $CH_3(CH_2)_7OOC\text{-}C_6H_4\text{-}COO(CH_2)_9CH_3$ | 119-07-3 |
| Diisononyl phthalate | DINP, DiNP | $C_6H_4[COO(CH_2)_6CH(CH_3)_2]_2$ | 28553-12-0 |
| Di(2-propylheptyl) phthalate | DPHP | $C_6H_4[COOCH_2CH(CH_2CH_2CH_3)(CH_2)_4CH_3]_2$ | 53306-54-0 |
| Diisodecyl phthalate | DIDP, DiDP | $C_6H_4[COO(CH_2)_7CH(CH_3)_2]_2$ | 26761-40-0 |
| Diundecyl phthalate | DUP | $C_6H_4[COO(CH_2)_{10}CH_3]_2$ | 3648-20-2 |
| Diisoundecyl phthalate | DIUP | $C_6H_4[COO(CH_2)_8CH(CH_3)_2]_2$ | 85507-79-5 |
| Ditridecyl phthalate | DTDP | $C_6H_4[COO(CH_2)_{12}CH_3]_2$ | 119-06-2 |
| Diisotridecyl phthalate | DITDP | $C_6H_4[COO(CH_2)_{10}CH(CH_3)_2]_2$ | 68515-47-9 |

Additionally, phthalates can leach, migrate, or evaporate into foodstuffs and beverages from packaging materials, so human exposure to phthalate esters is widespread [2, 12]. Leaked PAEs in plastic-packed foods can cause adverse effects on human health. Several PAEs are of great concern since they have been repeatedly detected in food and aqueous matrices due to migration from containers [13].

For instance, butyl benzyl phthalate (BBP) is used in flexographic inks for food packaging applications and it is considered to be a very toxic compound due to its mutagenic properties, acute oral and reproductive toxicity, and carcinogenicity [14, 15].

Based on several toxicological studies, some PAEs have been classified by their impact strength. For instance, Di(2-ethylhexyl) phthalate (DEHP) was in Class B2 and has been shown to be embryotoxic and teratogenic, while BBP was in Class C (possible human carcinogen). Di-n-butyl phthalate (DBP), diethyl phthalate (DEP) and dimethyl phthalate (DMP) were in class D (not yet classified as human carcinogens) [16-18]. Phthalates are also classified as endocrine disrupting chemicals (EDCs). There has been debate for and against the occurrence of EDCs in humans. However, there are increased incidences of abnormalities in human sexual and cognitive development in some societies with speculation that there are being caused by EDC exposure [17, 19, 20]. These abnormalities include low sperm count and decrease in sperm quality [21], and immunological and neurological effects. Considering that many of these compounds can elicit estrogenic responses at very low concentrations (parts per billion to parts per trillion), there is need for concern as many of the phthalates have been found at measurable concentrations in wastewater, surface waters, sediments, groundwater, and even drinking water in many countries [22-27].

Since 14 December 2005, the EU Directive 2005/84/EC [28] has banned several phthalates for use in PVC and other plasticized materials in all toys and childcare articles. Moreover, several studies have concluded that diet is the major route of exposure for humans, and that environmental contamination is one of the sources of these chemicals in food at various levels. So, the development of analytical methodologies for the monitoring of these compounds in a great variety of matrices including environmental ones, food products and food packaging materials, as well as in biological fluids for epidemiological studies on human exposure is mandatory.

Nowadays, modern analytical instrumentation has provided an edge to fast and cost-effective analytical methods [29, 30]. Nevertheless, many of these new analytical techniques have commonly suffered from some limitations such as the sample matrix effect and compromised selectivity in the analysis of complicated samples [31]. So, frequently, an effective sample clean-up helps to improve separation and detection, while a poor pretreatment procedure might invalidate the whole analysis. Furthermore, cleaned samples extend the service life of analytical instrumentation and reduce analytical cost [32]. Currently, numerous efficient and accurate sample preparation methods are widely used on PAE detection, such as solid-phase extraction (SPE) [33-35], solid-phase microextraction (SPME) [36], headspace solid-phase microextraction (HS-SPME) [37, 38], liquid-phase microextraction (LPME) [39], dispersive liquid-liquid microextraction (DLLME) [40, 41], molecularly imprinted polymers (MIP) sorbents [42], dispersive SPE (dSPE) [43], and even QuEChERS (*Quick, Easy, Cheap, Effective, Rugged and Safe*) procedures [44].

Most methods used for phthalates analysis are based on gas chromatography (GC) [45-48], including the EPA 606 method for organic chemical analysis of municipal and industrial wastewater [49]. Liquid chromatography (LC) has also been applied to this subject [45, 47]. LC techniques have several advantages over GC, i.e. less sample clean-up is required, thermally labile compounds are more easily analyzed, derivatization is usually not required and polar and high-molecular-weight compounds can be identified. So today, liquid chromatography coupled with mass spectrometry or tandem mass spectrometry (LC-MS(/MS)) is becoming a popular technique for the analysis of phthalates compounds in a great variety of samples [50, 51].

It should be noted that the determination of phthalates is not an easy task. Actually, due to the fact that these compounds are widely used, they are present in the laboratory environment, including air, and they can be absorbed into glass and other material, so false positive outputs can be produced during their analysis [52, 53]. Therefore, the risk of contamination is present in the whole analytical scheme, including sampling, sample preparation and analysis. This problem can be diminished using different methods proposed in the literature to prevent phthalate contamination problems [52] and by reducing the number of steps necessary to prepare the sample.

Nowadays, high resolution mass spectrometry (HRMS) and accurate mass measurements have recently gained great popularity in the determination of PAEs due to their capability to provide more comprehensive information concerning the exact molecular mass, elemental composition and detailed molecular structure of a compound [54-56]. Among the multiple benefits provided by HRMS over classical unit-mass-resolution tandem mass spectrometry we can find: (i) differentiation of isobaric compounds (different compounds with the same nominal mass but different elemental composition); (ii) simplification of sample-preparation procedures, thereby leading to faster methodologies requiring less sample manipulation; (iii) information gathered by a single injection that can be used for quantification and screening purposes, including targeted, suspect and non-targeted analysis; (iv) collection of full-scan spectra that can be stored and used in a later stage retrospective analysis, thus permitting the formulation of a posteriori hypotheses involving structural elucidation of unknown or suspected compounds [57, 58]. Moreover, HRMS appears as a helpful approach to prevent false positives in the analysis of phthalates.

## Sample Preparation Procedures

A wide variety of sample preparation has been reported in literature for phthalates. Typical steps in analytical procedures for sample preparation include sampling, homogenization, extraction, clean-up and concentration prior to instrumental analysis.

Regarding the extraction process, solvent extraction (SE) is probably the most common technique for extracting phthalates and metabolites from food, biological as well as environmental samples. Frequently, a mixture of solvents (e.g., dichloromethane with cyclohexane, acetonitrile–hexane, methanol–hexane–methyl tert-butyl ether, hexane–acetone and tetrahydrofuran–water) are employed depending on the physicochemical properties of the matrices [51]. In particular, special attention should be paid to samples containing lipids (i.e., fat foods like milk or meat). In fact, in such cases, liquid–liquid extraction inevitably co-extract lipids and organic compounds together with phthalates (because of the lipophilic properties of phthalates), thus demanding an additional clean-up step prior to LC analysis for the removal of lipids and other interferences from the extracts. Other extraction techniques like solid phase microextraction (SPME) [59] and pressurized liquid extraction (PLE) have also been used. For instance, Celeiro et al. [60] have recently developed a method based on PLE for the simultaneous determination of sixty-five chemical compounds (fragrance allergens, preservatives, musks, and phthalates) in wipes and wet toilet paper for children. Five phthalates (including the forbidden phthalates dibutyl phthalate and di(2-ethylhexyl)phthalate) were detected in the 18 analyzed samples.

Regarding the clean-up step, a sufficient degree of purification can be achieved by using different approaches like thin layer chromatography (TLC) [61], gel permeation chromatography (GPC) [62], and solid phase extraction (SPE) [63, 64], with the latter being the most popular technique. The selection of SPE cartridges with appropriate sorbents is crucial for efficient and selective extraction of phthalate and metabolites from samples. For instance, the Oasis HLB SPE sorbent, consisting of a hydrophilic-lipophilic balanced copolymer of [poly(divinylbenzene)-co-N-vinylpyrrolidone] have been successfully employed for the purification of urinary phthalate metabolites [65]. However, in some cases, more than one type of sorbents must be used to properly purified samples [66] and, due to the drawbacks of the commonly used SPE phases, in the last few years, there has also been an increasing number of "in-house" prepared sorbents, designed with the aim of improving already available stationary phases. These new materials try to fulfill the requirements according to present needs, such higher specific surface area, selectively towards the target analytes, easy manipulation allowing coupling on-line configurations and higher biocompatibility, with the overall objective of the enhancement of the efficiency of the extraction process [67]. A recent study used magnetic core mesoporous shell microspheres with C18-modified interior pore-walls for fast extraction and analysis of phthalates in water samples [68]. In this work, the novel sorbent was effectively applied to the extraction and enrichment of phthalates in water samples. In fact, thanks to the abundant exterior surface of silanol groups, the interior surface of functional C18 groups and the strong magnetic responsivity, the core–shell C18-functionalized $Fe_3O_4$@$mSiO_2$ microspheres showed good dispersibility and fast magnetic response, which made the process of organic compounds extraction very simple, quick and efficient. Previously, polypyrrole (PPy)-coated $Fe_3O_4$ magnetic microsphere were successfully synthesized, and applied as a magnetic sorbent to extract and concentrate phthalates in aquatic samples. In this study, seven kinds of phthalates were selected as model analytes, including dimethyl phthalate (DMP), diethyl phthalate (DEP), di-iso-butyl phthalate (DIBP), di-n-butyl phthalate (DBP), benzylbutyl phthalate (BBP), di-(2-ethylhexyl) phthalate (DEHP) and di-n-octyl phthalate (DNOP). Results from validation experiments showed good linearity (0.985–0.998), precision (3.4–11.7%), and high recovery (91.1–113.4%). More recently, Wang et al. [69] also developed a magnetic graphene @hydrophilic polydopamine for the enrichment and analysis of phthalates in environmental water samples.

Finally, one of the most important problems in phthalates analysis is the contamination problem, resulting in both false positive results or over estimated levels. The "blank" problem lies in the fact that these compounds are inherently ubiquitous in the laboratory environment, and they can be introduced into the sample during sample treatment [51]. Sources of phthalates in the laboratory environment were investigated by Fankhauser-Noti and Grob [52]. A 1.5-mL autosampler vial was shown to contain 10 ng of DBP and 4 ng of DEHP, whereas the concentrations of DBP and DEHP in the laboratory air were estimated to be 3 $\mu g/m^3$ and 2.4 $\mu g/m^3$, respectively. So, the risk of contamination is extremely high and it can only be reduced if the sample preparation is kept to a minimum, with minimal extraction steps, minimal extract concentration and minimal glassware use. Within this scenery, the use of dispersive SPE by QuEChERS for both rapid extraction and clean-up of phthalates may represents a valuable option to reduce blank problems. As an example, Jia et al. [56] developed an easy-to-use sample preparation based on QuEChERS for the simultaneous extraction of the 27 most important phthalates from milk (cow fat content > 2%), milk beverages (cow, protein content > 0.7%) and yogurt (cow, fat content > 3%). The proposed methodology showed extraction recoveries ranging from 90.7% to 104.6%, with coefficient of variation <5.6%.

## LIQUID CHROMATOGRAPHY-MASS SPECTROMETRY

Nowadays, liquid chromatography-mass spectrometry (LC-MS(/MS)) is becoming very popular for the determination of phthalates. Table 2 shows a selection of published LC-MS(/MS) methods for that purpose [24, 41, 43, 64, 70-81]. As can be seen, LC-MS(/MS) methods have been employed for determining phthalates in the analysis of a great variety of matrices, including environmental samples such as waters [74, 76], wastewaters [75] and sediments [24], biological fluids such plasma and feces [77, 79], pharmaceutical formulae and physiological solutions [70-73], cleaning and personal care products [41], food products and beverages [43, 76, 80, 81], food simulants from plastic food contact materials [64], and even on Children's toys [78], among others. Reversed-phase using C8- [43, 70, 71, 73, 76] or C18-bonded [24, 41, 64, 72, 74, 75, 77, 79, 80] stationary phases is typically employed for the LC-MS chromatographic separation of phthalates under isocratic or gradient elution conditions.

## Table 2. Selection of LC-MS(/MS) methods for the determination of phthalates

| Compounds | Samples | Sample treatment | Chromatographic conditions | Mass spectrometry conditions | Analysis time | LOQs | Ref. |
|---|---|---|---|---|---|---|---|
| DMP, DEP, DPP, BBP, DcHP | Fruit jellies | QuEChERS<br>- 10 mL acetonitrile<br>- 4 g $MgSO_4$ and 1 g NaCl<br>- 25 mg PSA and 200 mg $MgSO_4$ | Inertsil C8-3 column (150x2.1 mm, 5 µm)<br>*Gradient elution:*<br>A) Water<br>B) Methanol<br>0.4 mL/min | Single quadrupole mass analyzer<br>(+) ESI<br>SIM acquisition mode | 9 min | 0.28-11.25 µg/L | [43] |
| DMP, BBP, DEP, DBP, DPeP, DEHP, DOP | Pharmaceutical formulae used in parenteral nutrition | Liquid-liquid extraction with hexane after addition of 0.1 M NaOH | Zorbax Eclipse XDB-C8 column (50x2.1 mm, 3.5 µm)<br>*Gradient elution:*<br>A) Water (0.1% acetic acid)<br>B) Acetonitrile (0.1% acetic acid)<br>0.2 mL/min | Single quadrupole mass analyzer<br>(+) ESI<br>SIM acquisition mode | 30 min | 0.5-44.1 µg/L | [70] |
| DMP, DEP, BBP, DBP | Physiological saline solution | Direct analysis | Zorbax Eclipse XDB-C8 column (50x2.1 mm, 3.5 µm)<br>*Gradient elution:*<br>A) Water (0.1% acetic acid)<br>B) Acetonitrile (0.1% acetic acid)<br>0.2 mL/min | Triple quadrupole mass analyzer<br>(+) ESI<br>SRM acquisition mode | 12 min | 0.16-2.74 µg/L | [71] |
| DMP, DEP, DPrP, BBP, DcHP, DOP | Physiological saline solution | C18 Monolithic silica spin column extraction (MonoSpin-SPE)<br>-conditioning with methanol<br>-elution with acetone | Shim-pack XR-ODS II column (75x2.0 mm, 2.2 µm)<br>*Gradient elution:*<br>A) Water<br>B) Methanol<br>0.4 mL/min | Single quadrupole mass analyzer<br>(+) ESI<br>SIM acquisition mode | 9 min | 0.08-2.7 µg/L | [72] |
| DMP, DEP, BBP, DBP | Contact lens and cleaning solutions | Direct analysis<br><br>For contact lens, extraction with artificial tear solution and direct analysis | Zorbax Eclipse XDB-C8 column (50x2.1 mm, 3.5 µm)<br>*Gradient elution:*<br>A) Water (0.1% acetic acid)<br>B) Acetonitrile (0.1% acetic acid)<br>0.2 mL/min | Triple quadrupole mass analyzer<br>(+) ESI<br>SRM acquisition mode | 14 min | 0.09-8.74 µg/L | [73] |

| Compounds | Samples | Sample treatment | Chromatographic conditions | Mass spectrometry conditions | Analysis time | LOQs | Ref. |
|---|---|---|---|---|---|---|---|
| DEP, DBP, BBP, DEHP | Water | Direct analysis | Acquity UPLC BEH-C18 column (50x2.1 mm, 1.7 µm)<br>*Gradient elution:*<br>A) Water<br>B) Methanol<br>0.2 mL/min | Triple quadrupole mass analyzer<br>(+) ESI<br>SRM acquisition mode | 4 min | 0.06-0.54 µg/L | [74] |
| BBP with 5 other endocrine disrupter compounds | Wastewaters | Direct analysis | Zorbax SB-C18 column (100x2.1 mm, 3.5 µm)<br>*Gradient elution:*<br>A) Water (0.1% formic acid and 5 mM ammonium formate)<br>B) Methanol (0.1% formic acid and 5 mM ammonium formate)<br>0.2 mL/min | Triple quadrupole mass analyzer<br>(+) ESI<br>SRM acquisition mode | 9 min | 0.13-0.43 µg/L | [75] |
| 16 PAEs | Food simulants from plastic food contact materials | Direct analysis of food simulants:<br>- distilled water<br>- 3% acetic acid<br>- 10% alcohol<br>C18 SPE for olive oil simulant:<br>- elution with acetone:hexane (1:1 *v/v*) | Zorbax SB-C18 column (100x2.1 mm, 3.5 µm)<br>*Gradient elution:*<br>A) Water (10 mM ammonium acetate, pH 4.0)<br>B) Acetonitrile<br>0.4 mL/min | Triple quadrupole mass analyzer<br>(+) ESI<br>SRM acquisition mode | 18 min | 1.4-19.2 µg/kg<br>32.6-390.8 µg/kg (olive oil)<br>(LODs) | [64] |
| DEHP with bisphenol A and melamine | Soft drinks, water and dry milk powder | Water and soft drinks:<br>- LiChroprep RP-2 column and C18 column<br>- Extraction with 10 mL $CH_2Cl_2$ and elution with methanol (10% $NH_3$)<br>Dry milk powder:<br>- 10 mL methanol (10% $NH_3$) | Agilent Eclipse XDB-C8 column (150x4.6 mm, 5 µm)<br>*Gradient elution:*<br>A) Water (0.5% acetic acid and 0.1 mM sodium acetate)<br>B) Methanol:water (9:1 *v/v*) (0.5% acetic acid and 0.1 mM sodium acetate)<br>0.5 mL/min | Triple quadrupole mass analyzer<br>(+) ESI<br>SRM acquisition mode | 22 min | 0.046 µ/L | [76] |

**Table 2. (Continued)**

| Compounds | Samples | Sample treatment | Chromatographic conditions | Mass spectrometry conditions | Analysis time | LOQs | Ref. |
|---|---|---|---|---|---|---|---|
| | | - LiChroprep RP-2 column and C18 column<br>- Extraction with 10 mL $CH_2Cl_2$ and elution with methanol (10% $NH_3$) | | | | | |
| DBP | Rat plasma and feces | - Centrifugation of blood to obtain plasma<br>- Extraction of feces with acetonitrile<br>- Protein precipitation with methanol in both plasma and feces | Acquity UPLC BEH C18 column (100x2.1 mm, 1.7 µm)<br>*Isocratic elution:*<br>5 mM ammonium acetate – methanol (27:73 *v/v*)<br>0.25 mL/min | Triple quadrupole mass analyzer<br>(+) ESI<br>SRM acquisition mode | 6 min | 5 µg/L (plasma)<br>125 µg/kg (feces) | [77] |
| DBP, BBP, DEHP, DNOP, DINP, DIDP | Children's toys | - Extraction with methane chloride | XBridge C18 column (100x2.1mm, 3.5 µm)<br>*Gradient elution:*<br>A) Water (10 mM ammonium acetate)<br>B) Acetonitrile<br>0.5mL/min | Triple quadrupole mass analyzer<br>(+) ESI<br>SRM acquisition mode | 10 min | 0.96-4.97 µg/L | [78] |
| DEHP | Rat plasma and feces | Rat plasma<br>- Protein precipitation with methanol<br>Feces<br>- Extraction with acetonitrile and protein precipitation with methanol | Acquity UPLC BEH C18 column (100x2.1 mm, 1.7 µm)<br>*Isocratic elution:*<br>5 mM ammonium acetate-methanol (11:89 *v/v*)<br>0.25 mL/min | Triple quadrupole mass analyzer<br>(+) ESI<br>SRM acquisition mode | 6 min | 100 µg/L | [79] |

| Compounds | Samples | Sample treatment | Chromatographic conditions | Mass spectrometry conditions | Analysis time | LOQs | Ref. |
|---|---|---|---|---|---|---|---|
| 23 PAEs | milk-based products, distilled liquor, wine, beverage, grain, meat, oil, cookies and canned food | Liquid samples<br>Extraction with acetonitrile and saturation of water content with NaCl<br>Solid samples<br>QuEChERS<br>- 5 mL water and 15 mL acetonitrile<br>- 6g $MgSO_4$ and 1.5 g sodium acetate<br>- 50 mg PSA, 50 mg C18, 150 mg $MgSO_4$ | Poroshell 120 EC-C18 column (100x4.6 mm, 2.7 µm)<br>*Gradient elution:*<br>A) Water (0.1% formic acid)<br>B) Methanol<br>0.7 mL/min | Triple quadrupole mass analyzer<br>(+) ESI<br>SRM acquisition mode | 16 min | 10 -100 µg/kg | [80] |
| DMP, DEP, DiBP, DnBP, BBP, DEHP, DOP, DiNP, DiDP | Wine | - Dilution with methanol and direct analysis | Synergy Hydro-RP HPLC column (75x2.0 mm, 4 µm)<br>*Gradient elution:*<br>A) Water (10 mM ammonium acetate)<br>B) Methanol<br>0.3 mL/min | Triple quadrupole mass analyzer<br>(+) ESI<br>SRM acquisition mode | 20 min | 1.6-26.6 µg/L | [81] |
| DMP, DEP, BBP, DBP, DEHP, DOP | Cleaning and personal care products | - Ultrasound assisted extraction with acetonitrile<br>- Dispersive liquid-liquid microextraction (DLLME): 3 mL acetonitrile, 150 µL carbon tetrachloride, 10 mL aqueous solution | ODS column (150x4.0 mm, 4 µm)<br>*Gradient elution:*<br>A) 10 mM ammonium acetate (pH 4)<br>B) Acetonitrile<br>0.4 mL/min | Ion trap mass analyzer<br>(+) ESI<br>product ion scan (*m/z* 50-500)<br>SRM acquisition mode | 20 min | 40-450 µg/L (LODs) | [41] |
| BBP, DnBP, DiDP, DEHP, DiNP, DnOP, with several antibiotics | River sediments | - Extraction by sonication with methanol<br>- Clean-up with silica gel | Eclipse Plus C18 column (100x2.1 mm, 1.8 µm)<br>*Gradient elution:*<br>A) Water (0.1% formic acid)<br>B) Methanol<br>0.2-0.55 mL/min | Triple quadrupole mass analyzer<br>(+) ESI<br>SRM acquisition mode | 12 min | - | [24] |

Generally, water and methanol or water and acetonitrile mixtures, sometimes under acidic conditions with formic or acetic acids, are used as mobile phases, and in some cases other modifiers such as ammonium acetate or sodium acetate are also employed (see Table 2).

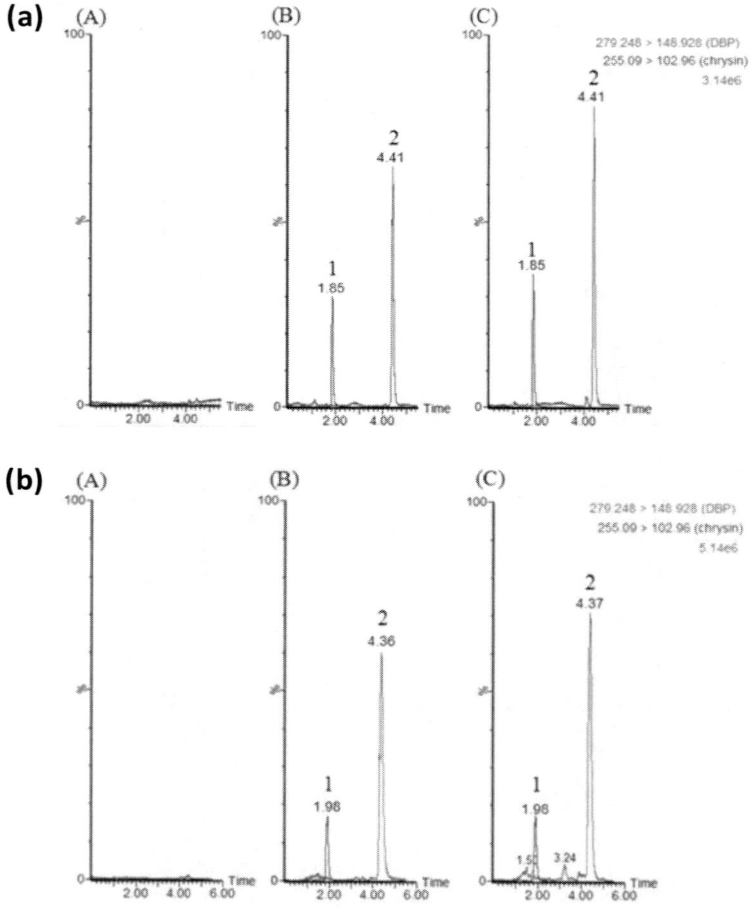


Figure 2. (a) Chromatograms of (A) blank plasma after protein precipitation, (B) blank plasma spiked with DBP (1 µg/mL) and chrysin (1 µg/mL) and (C) real sample collected at 5 min after DBP administration (30 mg/kg, i.v.) with chrysin (1 µg/mL). (b) Chromatograms of (A) blank feces after protein precipitation, (B) blank feces spiked with DBP (1 µg/g) and chrysin (1 µg/mL) and (C) real sample collected at 0-12 h after DBP administration (100 mg/kg, p.o.) with chrysin (1 µg/mL). Peak identification: 1, chrysin (used as internal standard); 2, DBP. Reproduced from *Open Access* Ref. [77].

Some of these works are focused in the determination of a specific phthalate compound, generally in the analysis of biological fluids to carry out pharmacokinetic studies.

For instance, Chang et al. [77] studied the pharmacokinetics of dibutyl phthalate in rat plasma and feces by UHPLC-MS/MS using sub-2μm particle size column (Acquity BEH C18 (100x2.1 mm, 1.7 μm)) and a 5 mM ammonium acetate-methanol (27:73 *v/v*) solution as mobile phase under isocratic conditions. Figure 2 shows the chromatograms of blank, spiked and real plasma (Figure 2a) and feces (Figure 2b) samples, respectively.

The UHPLC-MS/MS system equipped with positive electrospray ionization (ESI) source in multiple reaction monitoring (MRM) mode was used to monitor the *m/z* 279.25 → 184.93 selective reaction monitoring (SRM) transition for DBP. Limits of detection (LODs) in rat plasma and feces were 0.05 μg/mL and 0.125 μg/g, respectively, within an analysis time of 6 min.

The pharmacokinetic results demonstrated that DBP appeared to have a two-compartment model in the rats; the area under concentration *versus* time (AUC) was 57.8 ± 5.93 min μg/mL and the distribution and elimination half-life were 5.77 ± 1.14 and 217 ± 131 min, respectively, after DBP administration (30 mg/kg, i.v.). About 0.18% of the administered dose was recovered from the feces within 48 h.

This pharmacokinetic behavior demonstrated that DBP was quickly degraded within 2 h, suggesting a rapid metabolism with low fecal cumulative excretion in the rats.

In a similar work, Chang-Liao et al. [79] proposed the same UHPLC-MS/MS methodology to study the pharmacokinetics of di-(2-ethylhexyl) phthalate in rats, although a 5 mM ammonium acetate-methanol (11:89 *v/v*) solution was used as mobile phase. As an example, Figure 3 shows the representative UHPLC-MS/MS chromatograms obtained for both rat plasma and rat feces samples.

In this case, the SRM transitions monitored were *m/z* 391.4 → 149.0 for DEHP and *m/z* 313.3 → 149.0 for BBP. Although protein precipitation failed to completely remove the unwanted plasma components, the matrix effect of DEHP for both plasma and feces in the methanol extracts analyzed ranged from 88.5 to 99.8% and 98.3 to 102.7%, respectively, values that were considered acceptable by the authors. Recovery values of DEHP were between 87.4% and 104.8% with standard deviation less than 5.9% in biological samples.

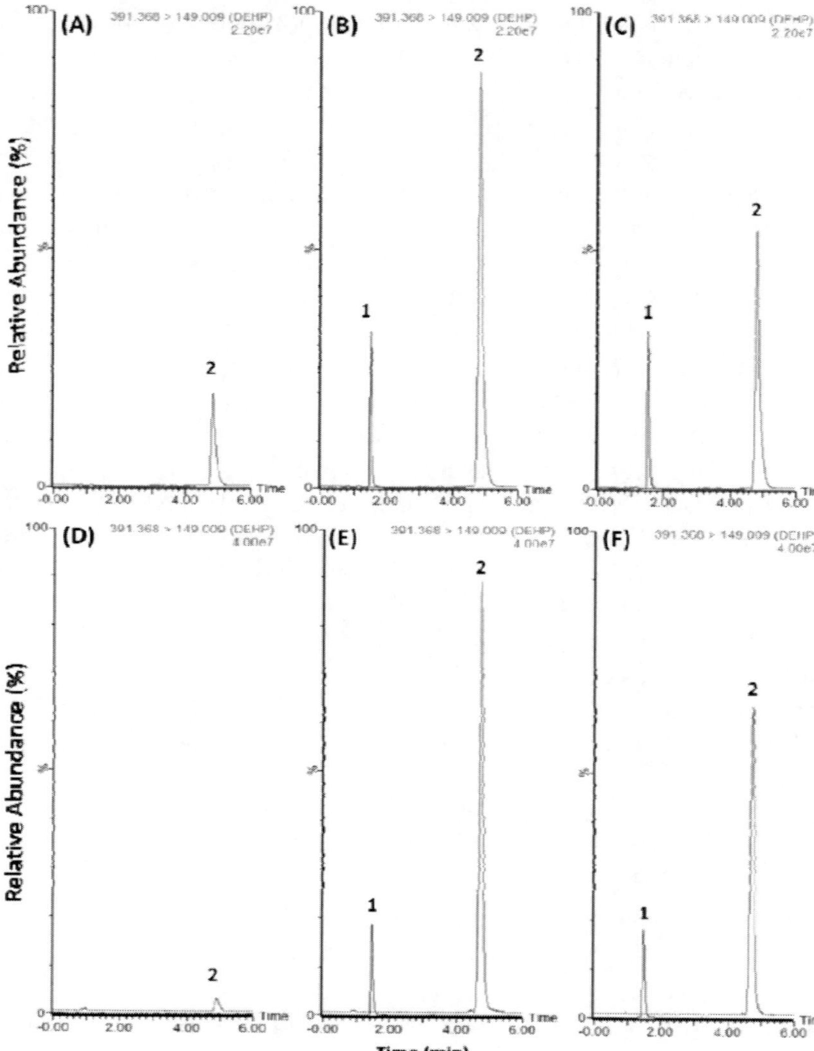


Figure 3. Representative UHPLC-MS/MS chromatograms of DEHP and BBP (used as internal standard). (A) Blank plasma sample; (B) blank plasma sample spiked with 0.5 μg/mL DEHP, and (C) plasma sample collected from rat plasma after administration of DEHP (10 mg/kg, i.v.). (D) Blanc fecal sample; (E) blank fecal sample spiked with 1 μg/mL DEHP, and (F) fecal sample (400-fold dilution) collected from rat feces after administration of DEHP (100 mg/kg, p.o.) at 0-12 h. Peak identification: 1, BBP; 2, DEHP. Reproduced from *Open Access* Ref. [79].

However, most of the LC-MS/MS methods reported in the literature are dealing with the analysis of multiple PAEs. For instance, recently Xu et al. [80] proposed the determination of 23 phthalates in food samples including milk-based products, distilled liquor, wine, beverages, grain, meat, oil, biscuit (cookie), and canned food by UHPLC-ESI-MS/MS using a fused-core C18 column (100x4.6 mm, 2.7 μm) and gradient elution with water (0.1% formic acid) and methanol as mobile phases. Separation of target phthalates was achieved within 16 min. In contrast to the two previously commented works, in this case acquisition was performed in SRM mode by monitoring two SRM transitions in order to comply with Commission Decision 657/2002/EC [82]. Figure 4 shows, as examples, the structures of the ions and fragmentation patterns corresponding to the SRM transitions selected for DBP and DEHP.

For instance, in the case of these two phthalates, Xu et al. proposed the monitoring of the transitions *m/z* 279.4 → 205.1 (DBP) and *m/z* 319.3 → 148.7 (DEHP) for quantitation purposes and the transitions *m/z* 279.4 → 149.2 (DBP) and *m/z* 319.3 → 113.0 (DEHP) for confirmation purposes [80]. LODs in the range 0.8-15 μg/kg were reported. The authors applied the proposed method to the analysis of 623 real samples collected from Taiwan Straits. DEHP was found in almost all the analyzed samples.

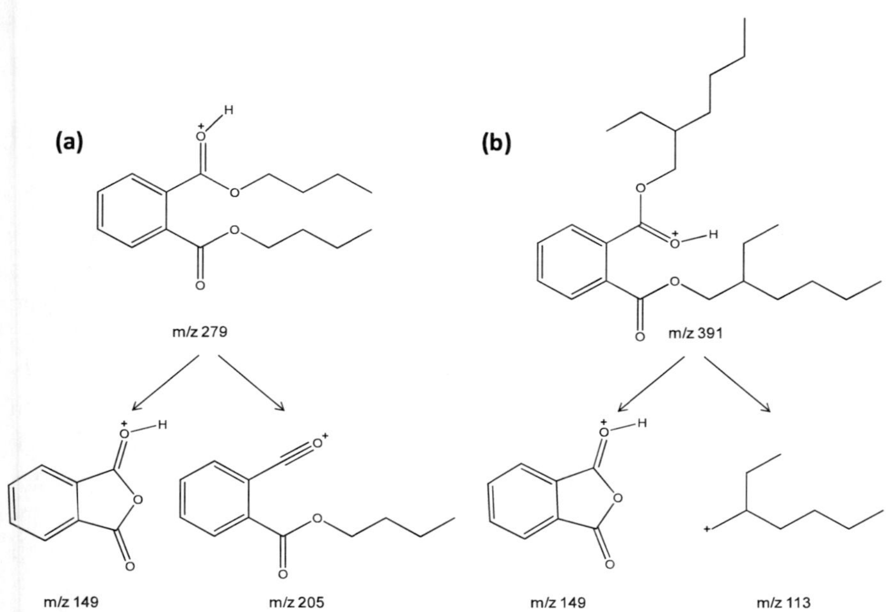


Figure 4. Examples of the fragmentation pathway of SRM transitions selected in the determination of (a) DBP and (b) DEHP.

Regarding mass spectrometric conditions, electrospray in positive ionization mode is always selected for the determination of PAEs, mainly using single quadrupole (for LC-MS) [43, 70, 72] and triple quadrupole (for LC-MS/MS) [24, 64, 71, 73-81] mass analyzers. However, some works using other low resolution mass analyzers can also be found in the literature. For instance, recently Viñas et al. [41] reported the application of an LC-ESI-MS/MS method using an ion-trap mass analyzer for the monitoring and determination of six phthalates (DMP, DEP, BBP, DBP, DEHP and DOP) in cleaning and personal care products. Separation was achieved with a conventional C18 reversed-phase column (150x4.0 mm, 4 μm) and an analysis time of 20 min. In this case, acquisition was carried out in product ion scan mode (*m/z* 50-500), being able to obtain the full MS/MS spectra. However, for quantitation and confirmation purposes two (for DEP) or three (for the other five targeted phthalates) SRM transitions were used. For example, for DBP SRM transitions corresponding to *m/z* 279 → 205, 279 → 149 and 279 → 91 were monitored. Quantification was performed by matrix-matched calibration. Detection limits in the range 0.04-0.45 ng/mL for the six PAEs considered were achieved, with recoveries between 84-124% and RSDs lower than 10%. The authors applied the proposed LC-ESI-MS/MS method for the analysis of thirty three samples including glass cleaners, degreasers, dishwashing and laundry detergents, and softeners, as well as personal care products such as shampoo, foot cream, deodorant, body milk and make-up, among others. Diethyl phthalate was the most frequently found compound at concentrations ranging from 0.154 to 81.663 μg/kg.

Because of the importance of the phthalate family of compounds in both environmental and food fields they are usually analyzed together with other contaminants. For instance, Petrovic and Barceló [83] proposed several years ago an LC-MS method for the determination of anionic and nonionic surfactants, their degradation products, and endocrine-disrupting compounds in sewage sludge. Among the analyzed compounds, three phthalates (DEP, DBP and DEHP) were targeted. In contrast to the most frequently ionization source used for the determination of phthalate compounds (ESI), in this work the determination of less polar compounds such as alcohol ethoxylates (AEOs), nonylphenol ethoxylates (NPEOs), coconut diethanol amides, poly(ethylene glycol)s, and phthalate esters was accomplished by reversed-phase liquid chromatography-atmospheric pressure chemical ionization-mass spectrometry (LC-APCI-MS) in positive ionization mode and with a conventional C18 column (chromatogram in Figure 5A).

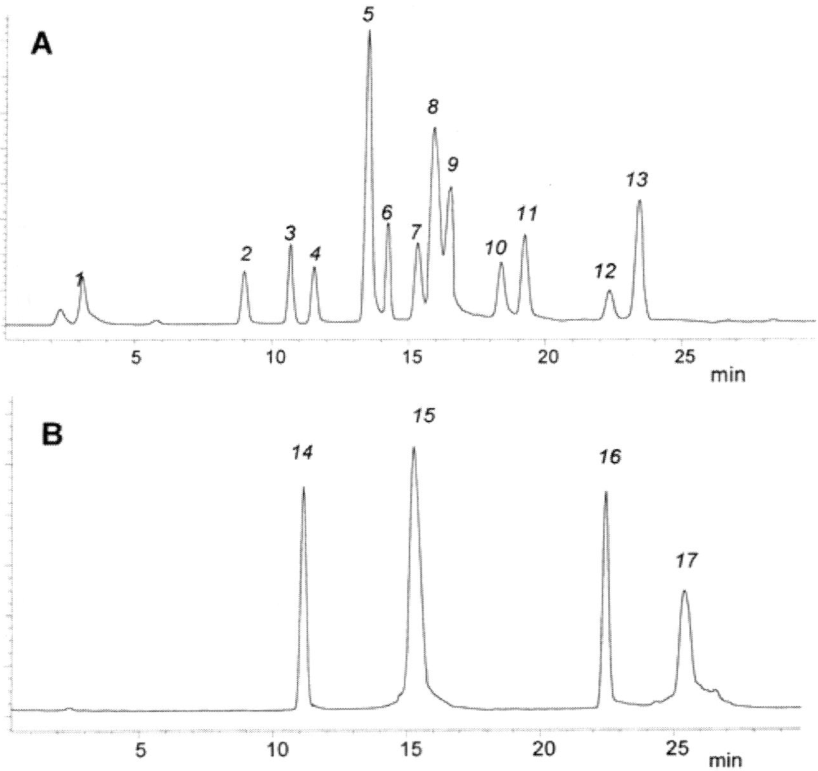


Figure 5. LC-MS chromatograms obtained by injecting the composite working standard solution containing target compounds. (A) Total ion chromatogram (TIC) obtained using APCI-MS detection in positive ion mode; (B) TIC chromatogram obtained using ESI-MS detection in negative ion mode. Peak identification: 1, polyethylene glycol; 2, $C_7$-coconut diethanol amide (CDEA); 3, DEP; 4, $C_9$-CDEA; 5, $C_{11}$-CDEA; 6, DBP; 7, $C_{10}$-alcohol ethoxylate (EO); 8, $C_{13}$-CDEA; 9, nonylphenol ethoxylate; 10, $C_{12}$-EO; 11, $C_{15}$-CDEA; 12, $C_{14}$-EO; 13, DEHP; 14, bisphenol A (BPA); 15, nonylphenol ethoxycarboxylate; 16, 4-octlphenol; and 17, 4-nonylphenol. Reprinted with permission from reference [83]. Copyright (2000) American Chemical Society.

More polar compounds such as nonylphenolcarboxylates, nonylphenol (NP), octylphenol, and bisphenol A (BPA) were analyzed by ion-pair LC-ESI-MS under negative ionization conditions (chromatogram in Figure 5B). LODs in the range 15-50 ng/g were established for the three analyzed phthalates. The authors applied the proposed method to the trace determination of targeted compounds in sewage sludge collected from different sewage treatment plants.

Reported levels of phthalates in the analyzed sludges were 0.2-1.9 mg/kg, 0.25-9.7 mg/kg, and 8-27 mg/kg for DEP, DBP and DEHP, respectively.

Recently, Komesli et al. [75] described an analytical procedure for the simultaneous determination of six selected endocrine disrupter compounds (diltiazem, progesterone, BBP, estrone, carbamazepine, and acetaminophen) by LC-ESI-MS/MS in wastewater samples. BBP was detected in all the influent and effluent waters of the treatment plants analyzed at concentrations ranging 0.06-0.18 μg/L.

In a more recent work, DEHP was determined together with bisphenol A and melamine in soft drinks, water and milk powder samples by LC-MS [76]. The author proposed one extraction procedure combining two extraction columns, a silica gel LiChroprep RP-2 column (20:1, g/g, top column) and a Sep-Pak with a C18 column (500 mg, bottom column) to optimally recover all three targeted analytes from the different matrices.

Chromatographic separation was achieved by using a conventional C8 column (150x4.6 mm, 5 μm) and using acidic water and methanol mobile phases. LODs for DEHP of 13 ng/L in water and soft drink samples and 0.65 μg/kg in dry milk powder were obtained. Concentrations levels in the range 0.03-9.05 ng/L in waters and soft drink samples, and from 5,300 and up to 25,100 ng/kg in milk powder samples were reported.

So far ingestion, inhalation, dermal absorption, and intravenous injection have been estimated as the main human exposure pathways to phthalate contaminants [84]. Moreover, medications could be another potential source of human exposure to these compounds [6, 85].

Once absorbed, phthalates are rapidly metabolized into their respective hydrolytic monoesters by hydrolysis, and then high molecular-weight monoesters (C>8) are further metabolized into more hydrophilic oxidative metabolites after enzymatic oxidation [86-88].

One way of monitoring human exposure to this kind of contaminants is through the analysis of phthalate metabolites in biological fluids such as human urine [63, 89-98], serum [90, 92], seminal plasma [90] and even in human hair [99].

Table 3 shows a selection of LC-MS/MS methods for the determination of phthalate metabolites in human biological samples. In contrast to the determination of PAEs by LC-ESI-MS/MS, the determination of phthalate metabolites is usually carried out by ESI in negative ionization mode and, in most of the cases, triple quadrupole or triple quadrupole/linear ion trap mass analyzers are usually employed.

**Table 3. Selection of LC-MS(/MS) methods for the determination of phthalate metabolites in human biological samples**

| Compounds | Samples | Sample treatment | Chromatographic conditions | Mass spectrometry conditions | Analysis time | LOQs | Ref. |
|---|---|---|---|---|---|---|---|
| 16 phthalate metabolites | Human urine | - automatic enzymatic hydrolysis<br>- on-line SPE with Chromolith Flash RP-18e precolumn (25x4.6 mm) | Betasil Phenyl analytical column (150x2.1 mm, 3 μm)<br>Gradient elution:<br>A) Water (0.1% acetic acid)<br>B) Acetonitrile (0.1% acetic acid)<br>0.3 mL/min | Triple quadrupole mass analyzer<br>(-) ESI<br>SRM acquisition mode | 27 min | 0.11-4.30<br>μg/L | [89] |
| 13 phthalate metabolites | Human urine, serum and seminal plasma | - Enzymatic hydrolysis (β-glucuronidase)<br>- Purification with Strata XS SPE cartridges (elution with acetonitrile and ethyl acetate) | Synergi 4U fusion-RP column (75x2.0 mm, 4 μm)<br>*Gradient elution:*<br>A) Water (0.1% acetic acid)<br>B) Acetonitrile (0.1% acetic acid)<br>0.3 mL/min | Triple quadrupole mass analyzer<br>(-) ESI<br>SRM acquisition mode | 24 min | 0.11-1.43<br>(urine)<br>0.10-0.92<br>(serum)<br>0.19-1.01<br>(seminal plasma)<br>μg/L<br>(LODs) | [90] |
| DEHP metabolites | Human urine | - Enzymatic hydrolysis (β-glucuronidase) | Gemini C6-phenyl column (100x2.0 mm, 3 μm)<br>*Gradient elution:*<br>A) 5 mM ammonium acetate buffer with 0.1% acetic acid (pH 3.5)<br>B) Acetonitrile<br>0.25 mL/min | Triple quadrupole/linear ion trap mass analyzer<br>(-) ESI<br>SRM acquisition mode | 12 min | 1.0<br>μg/L | [91] |
| DBP and DEHP metabolites | Human serum and urine | - Enzymatic hydrolysis (β-glucuronidase)<br>- Addition of 1 M ammonium acetate | Imtackt Cadenza CD C18 column (75x2.0 mm, 3 μm)<br>*Gradient elution:*<br>A) Water (0.1% formic acid)<br>B) Acetonitrile (0.1% formic acid)<br>0.6 mL/min | Triple quadrupole mass analyzer<br>(-) ESI<br>SRM acquisition mode | 15 min | 0.23-2.02<br>(serum)<br>0.46-3.15<br>(urine)<br>μg/L | [92] |

**Table 3. (Continued)**

| Compounds | Samples | Sample treatment | Chromatographic conditions | Mass spectrometry conditions | Analysis time | LOQs | Ref. |
|---|---|---|---|---|---|---|---|
| DEHP and DiNP metabolites | Human urine | - Enzymatic hydrolysis (β-glucuronidase)<br>- Purification with Strata XS SPE cartridges (elution with acetonitrile and ethyl acetate) | Synergi 4U fusion-RP column (75x2.0 mm, 4 μm)<br>*Gradient elution:*<br>A) Water (0.1% acetic acid)<br>B) Acetonitrile (0.1% acetic acid)<br>0.3 mL/min | Triple quadrupole mass analyzer<br>(-) ESI<br>SRM acquisition mode | 17 min | 0.10-3.94 μg/L (LODs) | [93] |
| DPHP metabolites | Human urine | - Enzymatic hydrolysis (β-glucuronidase)<br>- Addition of 1 M ammonium acetate | Zorbax Eclipse Plus C18 column (150x2.1 mm, 3.5 μm)<br>*Gradient elution:*<br>A) Water (0.1% formic acid)<br>B) Methanol<br>0.2 mL/min | Triple quadrupole mass analyzer<br>(-) ESI<br>SRM acquisition mode | 21 min | 0.1-0.2 μg/L (LODs) | [94] |
| Phthalate metabolites with parabens and benzophenone-3 | Human Urine | - Hydrolysis with acidified buffers and enzymatic treatments<br>- SPE Bond Elut Certify LRC cartridges<br>- Elution with acetonitrile | Kinetex Phenyl-Hexyl column (100x2.1 mm, 1.7 μm)<br>*Gradient elution:*<br>A) Water (0.1% acetic acid)<br>B) Acetonitrile (0.1% acetic acid)<br>0.55 mL/min | Triple quadrupole mass analyzer<br>(+) ESI<br>SRM acquisition mode | 14 min | 0.43-1.23 μg/L | [95] |
| 14 phthalate metabolites, bisphenol A and endogenous steroid hormones | Human urine | - Enzymatic hydrolysis (β-glucuronidase)<br>- Purification with Oasis MAX mixed-mode SPE cartridges (elution with methanol containing 1% formic acid) | Zorbax Eclipse Plus C18 column (150x2.1 mm, 3.5 μm)<br>*Gradient elution:*<br>A) Water (0.1% acetic acid)<br>B) Acetonitrile (0.1% acetic acid)<br>0.3 mL/min | Triple quadrupole mass analyzer<br>(-) ESI<br>SRM acquisition mode | 12 min | 0.10-2.80 μg/L | [96] |

| Compounds | Samples | Sample treatment | Chromatographic conditions | Mass spectrometry conditions | Analysis time | LOQs | Ref. |
|---|---|---|---|---|---|---|---|
| 5 monoester phthalate metabolites and bisphenol A | Human urine | - Enzymatic hydrolysis (β-glucuronidase)<br>- Addition of ammonium acetate buffer<br>- Purification with silica-based bonded C18 SPE cartridges (elution with acetonitrile) | Thermo Betasil phenyl column (50x2.1 mm, 3 μm)<br>*Gradient elution:*<br>A) Water (0.1% acetic acid)<br>B) Acetonitrile (0.1% acetic acid)<br>0.22 mL/min | Triple quadrupole mass analyzer<br>(-) ESI<br>SRM acquisition mode | 8 min | 0.3-1 μg/L | [97] |
| DEHP metabolites | Human hair | - Decontamination with dichloromethane<br>- Extraction with methanol/trifluoroacetic acid (8.5:1.5 *v/v*) | Luna C18 column (50x2.0 mm, 3 μm)<br>*Gradient elution:*<br>A) Water (0.1% formic acid)<br>B) Methanol (0.1% formic acid)<br>0.25 mL/min | Triple quadrupole mass analyzer<br>(-) ESI<br>SRM acquisition mode | 10 min | 1.0-5.0 ng/kg | [99] |
| 5 monoester phthalate metabolites | Human urine | - Enzymatic hydrolysis (β-glucuronidase)<br>- Addition of ammonium acetate buffer<br>- Purification with Agilent Bond Elut NEXUS SPE cartridges (elution with acetonitrile and ethyl acetate) | Agilent Eclipse Plus C18 column (100x2.1 mm, 3.5 μm)<br>*Gradient elution:*<br>A) Water (0.1% acetic acid)<br>B) Methanol (0.1% acetic acid)<br>0.25 mL/min | Triple quadrupole mass analyzer<br>(-) ESI<br>SRM acquisition mode | 15 min | 0.1-0.5 μg/L | [98] |
| Phthalate metabolites with bisphenol A and structural isomers of parabens | Human urine | - Enzymatic hydrolysis (β-glucuronidase)<br>- Addition of ammonium acetate buffer<br>- Purification with Agilent Bond Elut NEXUS SPE cartridges (elution with acetonitrile and ethyl acetate) | Thermo Betasil phenyl column (100x2.1 mm, 3 μm)<br>*Gradient elution:*<br>A) Water (0.1% acetic acid)<br>B) Acetonitrile (0.1% acetic acid)<br>0.35 mL/min | Triple quadrupole mass analyzer<br>(-) ESI<br>SRM acquisition mode | 35 min | 0.03-2.80 μg/L | [63] |

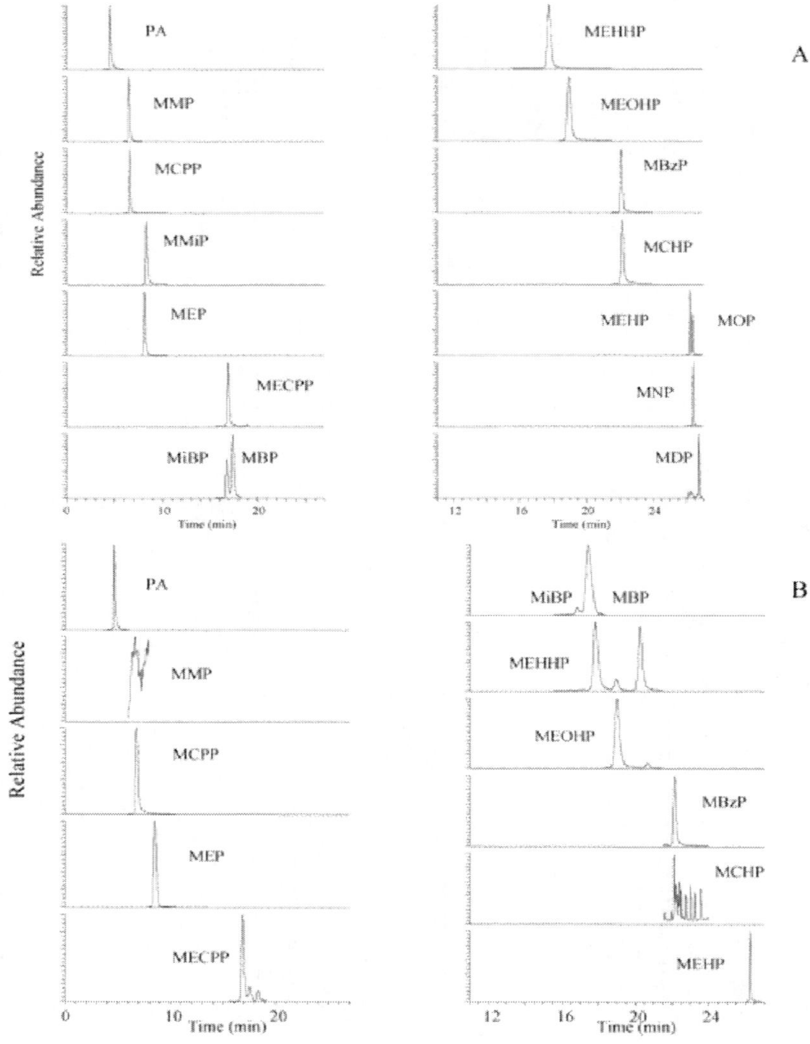


Figure 6. LC-ESI-MS/MS chromatogram of (A) a standard mixture (concentrations are between 30 and 35 ng/mL) and (B) a non-spiked human urine extract. PA: phthalic acid; MMP: monomethyl phthalate; MCPP: mono(3-carboxypropyl) phthalate; MMiP: monomethyl isophthalate; MEP: monoethyl phthalate; MECPP: mono(2ethyl-5-carboxypentyl) phthalate; MiBP: monoisobutyl phthalate; MBP: mono-n-butyl phthalate; MEHHP: mono(2-ethyl-5-hydroxyhexyl) phthalate; MEOHP: mono(2-ethyl-5-oxohexyl) phthalate; MBzP: monobenzyl phthalate; MCHP: monocyclohexyl phthalate; MEHP: mono(2-ethylhexyl) phthalate; MOP: mono-n-octyl phthalate; MNP: mono-iso-nonyl phthalate; and MDP: mono-iso-decyl phthalate. Reprinted from reference [89]. Not subject to U.S. Copyright. Publ. 2005 American Chemical Society.

Reversed-phase chromatography with C18 [90, 92-94, 96, 98, 99] columns are usually employed for the chromatographic separation, although other reversed-phase stationary phases such as phenyl-bonded phases have also been described [63, 89, 91, 95, 97].

For instance, Kato et al. [89] developed a method for the determination of 16 phthalate metabolites in human urine using an automated sample preparation procedure and on-line solid-phase extraction (SPE) preconcentration coupled with LC-ESI-MS/MS. Chromatographic separation was achieved by employing a Betasil Phenyl (150x2.1 mm, 3 μm) column. The obtained chromatograms for a standard mixture and a non-spiked human urine extract are shown in Figure 6.

In this work, the authors employed a complete automatic sample treatment procedure including spiking the urine sample with isotopically labeled phthalate metabolites, and the addition of 4-methylumbelliferone glucuronide and an enzyme solution to perform enzymatic hydrolysis. The same automatic procedure stopped the deglucuronidation reaction by addition of acetic acid and dilution with an acetonitrile:water solution. Then the sample was transferred to an on-line SPE-LC-ESI-MS/MS system for the preconcentration and determination of targeted phthalate metabolites. For SPE, a Chromolith Flash RP-18e monolithic column was employed. MRM was employed by monitoring two SRM transitions for quantitation and confirmation purposes. LODs between 0.11 and 4.30 ng/mL were achieved. This method was not labor intensive and, because pretreatment of the urine samples was performed automatically using an HPLC autosampler, involved minimal sample handling, thus minimizing exposure to hazardous chemicals. The method was validated by the authors on spiked, pooled urine samples and on urine samples from 43 adults with no known exposure to phthalates. Several compounds such as PA, MEP, MiBP, MBP, MECPP, MEOHP and MEHHP (chemical names in Figure 6 caption) were detected in all the human urine samples analyzed. The analytes were detected at a wide range of concentrations, confirming that human exposure to phthalates is widespread.

An interesting work is the one described by Chang et al. [99] for the determination of DEHP metabolites (mono-(2-ethylhexyl) phthalate, mono-(2-ethyl-5-hydroxyhexyl) phthalate, mono-(2-ethyl-5-oxy-hexyl) phthalate, mono-(2-ethyl-5-carboxypentyl) phthalate, and mono-[2-(carboxymethyl) hexyl] phthalate) in human hair using LC-ESI-MS/MS using a triple quadrupole mass analyzer. Separation was carried out by employing a conventional C18 reversed-phase column and acetic acid-ammonium acetate buffer solution and acetonitrile as mobile phases. Hair samples were collected at the posterior vertex of the head from ten individuals. After decontamination of hair samples with dichloromethane, samples were incubated in methanol:trifluoroacetic acid solution (8.5:1.5 *v/v)* for analyte extraction. DEHP metabolites were detected in human hair samples at concentrations ranging from 2.3 to 96.2 ng/kg, being mono-(2-ethylhexyl) phthalate the most abundant one, found at a mean value of 44.9 ng/kg.

## LIQUID CHROMATOGRAHPY-HIGH RESOLUTION MASS SPECTROMETRY

Considerable interest has recently arisen in emerging accurate-mass, high-resolution MS (HRMS) techniques. The high resolving power attainable with time-of-flight (TOF) mass analyzers and especially with Orbitrap instruments, the accurate mass measurements obtained with this kind of MS analyzers, and the availability of multiple searching engines such as mass spectra libraries are important tools to achieve identification and confirmation of target and non-target compounds in a great variety of matrices.

Regarding phthalates analysis, although many papers have now been published on LC-MS, GC-MS probably continued to be the technique of choice when performing phthalates and metabolites studies. And within LC-MS scenery, only few papers have been reported dealing with the use of HRMS.

Recently, an analytical strategy including both the quantitative target analysis of 8 regulated primary aromatic amines (PAAs), as well as a comprehensive post-run target screening of 77 migrating substances (including phthalates), has been developed for nylon utensils, using Orbitrap mass spectrometer operating in full scan mode [100]. The accurate mass data were acquired with a resolving power of 50,000 FWHM [scan range 80-800 Da; scan speed, 0.5 s (2 Hz)], and by alternating two acquisition events, ESI(+) with and without fragmentation. Extracted ion chromatograms (XIC) for individual compounds were reconstructed from the full-scan data with a mass tolerance of 5 ppm. The strategy was applied to 10 real samples collected from different retailers in the Valencia Region (Spain) during 2014 and, as result, di-n-butylphthalate and di-2-ethylhexyl phthalate substances, which were not initially included in the target analysis, have been successfully identified and confirmed in the post-run target screening analysis.

Previously, ultrahigh-performance liquid chromatography and electrospray ionization quadrupole Orbitrap high-resolution mass spectrometry (UHPLC/ESI Q-Orbitrap) has also been used for the separation and detection of 27 phthalates in milk and milk products [56].

The use of QuEChERS for phthalate extraction and clean-up allows to obtain good recoveries (90.7–104.6%) also in samples like milk and milk-based products that are considered complicated matrices because of the lipophilic properties of most phthalates. The mass resolution was set at 17,500 FWHM (*m/z* 200) with normalized collision energy (NCE) to obtain product ion spectra at 35%. Using this strategy, no overlapping in masses was observed between analytes and matrix components since coeluting matrix compounds from the matrix or noisy peaks were easily excluded, thus facilitating the identification and quantification of known or unknown analytes in a single run analysis. Furthermore, as highlighted by the authors, the use of characteristic fragments allowed to overcome the problems related to the presence of some co-eluting structural analogs and multiple isomers. For instance, DBP and DIBP are co-eluting multiple isomers but unequivocal fragments (*m/z* 223.09703, 205.08647, 95.04969, 59.04914) have been selected for their confirmation. Furthermore, in order to minimize the probability of false positives, mass extraction windows of 3.0 ppm were employed and non-negative samples were also carried out using isotopic patterns.

The results of the study showed that DEHP was present in almost all milk, milk beverages and yogurt samples tested in this study with levels ranging from 1 to 936 μg/kg and that 55% (53/96) of the samples were contaminated with two or more phthalates. In pasteurized and homogenized milk samples, DEP, BBP, and DEHP were detected quite frequently with levels over 10 μg/kg.

LC-TOF/MS has also been used for phthalate analysis. For instance, Dirwono et al. [101] developed a LC-TOF/MS method for the determination of 12 phthalates (diethyl phthalate, DEP; dipropyl phthalate, DPrP; di-n-butyl phthalate, DBP; diisobutyl phthalate, DIBP; di-n-pentyl phthalate, DPP; dicyclohexyl phthalate, DCHP; di-n-hexyl phthalate, DHP; benzylbutyl phthalate, BBP; diethylhexyl phthalate, DEHP; diisononyl phthalate, DINP; diisodecyl phthalate, DIDP) and one adipate (di ethylhexyl adipate, DEHA) in edible salts (raw salts, refined salts, refined salts with additives and baked salts) available in Korean food markets.

The authors reported that DBP/DIBP and DEHP levels in refined salts were much lower than those in raw salt, whereas baked salts did not contain any phthalate at all, thus indicating that the refining process plays an important role to remove these contaminants.

Interestingly, Self et al. [102], reported a method based on direct analysis in real time (DART) ionization coupled to a Thermo Exactive Orbitrap mass spectrometer for the analysis of selected phthalic acid esters (benzyl butyl phthalate, dibutyl phthalate, diisobutyl phthalate, di-2-ethylhexyl phthalate, di-n-octyl phthalate, diisononyl phthalate, and diisodecyl phthalate) in food and nutraceutical products.

The Exactive mass spectrometer was operated in medium (10,000 FWHM) resolution in positive ion mode. The method has been proven to perform well for most beverage/food sample types at 0.5 mg/mL, and at 1.0 mg/mL for all food samples. In addition, the method is shown to capably analyze select nutraceutical sample types at levels of 50 mg/g or less of each of the seven phthalate analytes.

## Conclusion

The widespread use and the application of phthalates, anthropogenic chemicals that are broadly employed in many household and personal care products, as well as plasticizers in various domestic and industrial products, have resulted in the ubiquitous presence of these compounds in the environment. These compounds are known to be endocrine disruptors, and to have some carcinogenic, mutagenic and toxic effects in reproduction, so they are receiving considerable attention in political and scientific circles. Although the use of several phthalates has been banned by legislation, diet continues to be the major route of exposure of these compounds for humans, being environmental contamination one of the sources of these chemicals in food. For this reason, the development of analytical methodologies for the monitoring of these compounds is necessary not only in environmental matrices, but also in food products and food packaging materials, and in biological fluids for epidemiological studies on human exposure.

GC-MS and LC-MS(/MS) methods are the most common techniques used for that purpose. The present chapter has reviewed the role of liquid chromatography-mass spectrometry techniques in the detection, determination and confirmation of phthalates in the last years by means of relevant applications. Novel sample treatment procedures such as the use of in-house SPE sorbents using magnetic core mesoporous shell microspheres, or QuEChERS procedures have also been addressed. Regarding LC-MS techniques, LC-MS/MS using triple quadrupole instruments is the technique of choice for the analysis of phthalates, mainly using triple quadrupole mass analyzers. However, lately the use of high resolution mass spectrometry has gained interest also in this field because of the high resolving power attainable with time-of-flight and Orbitrap mass analyzers as well as the accurate mass measurements obtained with this kind of instrumentation, and sure the number of LC-HRMS applications in the analysis of phthalates compounds will increase in the near future.

## Acknowledgments

This work has been funded by the Spanish Ministry of Economy and Competitiveness under the project CTQ2012-30836, and from the Agency for Administration of University and Research Grants (Generalitat de Catalunya, Spain) under the project 2014 SGR-539.

## References

[1] Swan, S. H. (2008). Environmental phthalate exposure in relation to reproductive outcomes and other health endpoints in humans. *Environ. Res.*, 108, 177-184.

[2] Heudorf, U., Mersch-Sundermann, V. & Angerer, J. (2007). Phthalates: toxicology and exposure. *Int. J. Hyg. Environ. Health*, 210, 623-634.

[3] Staples, C. A., Peterson, D. R., Parkerton, T. F. & Adams, W. J. (1997). The environmental fate of phthalate esters: a literature review. *Chemosphere*, 35, 667-749.

[4] Xu, G., Li, F. & Wang, Q. (2008). Occurrence and degradation characteristics of dibutyl phthalate (DBP) and di-(2-ethylhexyl) phthalate (DEHP) in typical agricultural soils of China. *Sci. Total Environ.*, 393, 333-340.

[5] Sathyanarayana, S., Karr, C. J., Lozano, P., Brown, E., Calafat, A. M., Liu, F. & Swan, S. H. (2008). Baby care products: possible sources of infant phthalate exposure. *Pediatrics*, 121, e260-e268.

[6] Hauser, R., Duty, S., Godfrey-Bailey, L. & Calafat, A. M. (2004). Medications as a source of human exposure to phthalates. *Environ Health Perspect*, 112, 751-753.

[7] Schettler, T. (2006). Human exposure to phthalates via consumer products. *Int. J. Androl.*, 29, 134-139.

[8] Peijnenburg, W. J. G. M. & Struijs, J. (2006). Occurrence of phthalate esters in the environment of the Netherlands. *Ecotoxicol. Environ. Saf.*, 63, 204-215.

[9] Liang, D. W., Zhang, T., Fang, H. H. P. & He, J. (2008). Phthalates biodegradation in the environment. *Appl. Microbiol. Biotechnol.*, 80, 183-198.

[10] US-EPA, US Environmental Protection Agency. 2009. Integrated Risk Information System (IRIS); Available from: www.epa.gov/ncea/iris/index.html (Accessed May 5, 2015).

[11] EFSA, European Food Safety Authority, 2004. Statement on the Scientific Panel on Food Additives, Flavourings, Processing Aids and materials in Contact with Food on the reclassification of some phthalates for consistency with the new SCF guidelines for food contact materials. Available from; www.efsa.europa.eu/en/efsajournal/doc/1062.pdf (Accessed May 5, 2015).

[12] S. Lacorte, M. Cortina, A. Guart, A. Borrell. Food-Packaging Contaminants, Chapter 11, in: O. Núñez, H. Gallart-Ayala, C.P.B. Martins, P. Lucci (Eds), Fast Liquid Chromatography-Mass Spectrometry Methods in Food and Environmental Analysis, Imperial College Press, London, p. 421-484, 2015.

[13] Cirillo, T. & Cocchieri, R. A. (2013). Phthalates in foods. *Woodhead Publ. Ser. Food Sci., Technol. Nutr.*, 247, 334-366.

[14] M. Bolgar, J. Hubball, J. Groeger, S. Meronek, (2008). Handbook for the Chemical Analysis of Plastic and Polymer Additives, CRC Press, Boca Raton, FL, USA.

[15] ECHA (2015). European Chemical Agency website. Available at: http://echa.europa.eu/ (Accessed May 5, 2015).

[16] Alatriste-Mondragon, F., Iranpour, R. & Ahring, B. K. (2003). Toxicity of di-(2-ethylhexyl) phthalate on the anaerobic digestion of wastewater sludge. *Water Res.*, 37, 1260-1269.

[17] Olujimi, O. O., Fatoki, O. S., Odendaal, J. P. & Okonkwo, J. O. (2010). Endocrine disrupting chemicals (phenol and phthalates) in the South African environment: a need for more monitoring. *Water SA*, 36, 671-682.

[18] Latini, G., Wittassek, M., Del Vecchio, A., Presta, G., De Felice, C. & Angerer, J. (2009). Lactational exposure to phthalates in Southern Italy. *Environ. Int.*, 35, 236-239.

[19] Campbell, C. G., Borglin, S. E., Green, F. B., Grayson, A., Wozei, E. & Stringfellow, W. T. (2006). Biologically directed environmental monitoring, fate, and transport of estrogenic endocrine disrupting compounds in water: A review. *Chemosphere*, 65, 1265-1280.

[20] Falconer, I. R., Chapman, H. F., Moore, M. R. & Ranmuthugala, G. (2006). Endocrine-disrupting compounds: a review of their challenge to sustainable and safe water supply and water reuse. *Environ. Toxicol.*, 21, 181-191.

[21] van Wyk, J. H., Pool, E. J. & Leslie, A. J. (2003). The Effects of Anti-Androgenic and Estrogenic Disrupting Contaminants on Breeding Gland (Nuptial Pad) Morphology, Plasma Testosterone Levels, and Plasma Vitellogenin Levels in Male Xenopus laevis (African Clawed Frog). *Arch. Environ. Contam. Toxicol.*, 44, 247-256.

[22] Cortazar, E., Bartolome, L., Delgado, A., Etxebarria, N., Fernandez, L. A., Usobiaga, A. & Zuloaga, O. (2005). Optimization of microwave-assisted extraction for the determination of nonylphenols and phthalate esters in sediment samples and comparison with pressurized solvent extraction. *Anal. Chim. Acta*, 534, 247-254.

[23] Adeogun, A. O., Richard, I. O., Dami, O. E., Azubuike, C. V., Adegbola, R. A., Adewuyi, G. A. & Arukwe, A. (2015). Environmental occurrence and biota concentration of phthalate esters in Epe and Lagos lagoons, Nigeria. *Mar. Environ. Res.*, Ahead.

[24] Yang, G. C. C., Wang, C. L. & Chiu, Y. H. (2015). Occurrence and distribution of phthalate esters and pharmaceuticals in Taiwan river sediments. *J. Soils Sediments*, 15, 198-210.

[25] Hassanzadeh, N., Esmaili Sari, A., Khodabandeh, S. & Bahramifar, N. (2014). Occurrence and distribution of two phthalate esters in the sediments of the Anzali wetlands on the coast of the Caspian Sea (Iran). *Mar. Pollut. Bull.*, 89, 128-135.

[26] Liu, X., Shi, J., Bo, T., Li, H. & Crittenden, J. C. (2015). Occurrence and risk assessment of selected phthalates in drinking water from waterworks in China. *Environ. Sci. Pollut. Res.*, Ahead.

[27] Domínguez-Morueco, N., González-Alonso, S. & Valcárcel, Y. (2014). Phthalate occurrence in rivers and tap water from central Spain. *Sci. Total Environ.*, 500-501, 139-146.

[28] Directive 2005/84/EC of the European Parliament and of the Council of 14 December 2005 amending for the 22nd time Council Directive 76/769/EEC on the approximation of the laws, regulations and administrative provisions of the Member States relating to restrictions on

the marketing and use of certain dangerous substances and preparations (phthalates in toys and childcare articles), *Official J. of the Eurpean Union* L344, 40-43.

[29] Núñez, O., Gallart-Ayala, H., Martins, C. P. B. & Lucci, P. (2012). New trends in fast liquid chromatography for food and environmental analysis. *J. Chromatogr. A*, 1228, 298-323.

[30] Núñez, O., Gallart-Ayala, H., Martins, C.P.B. & Lucci, P. (Eds) (2015), Fast Liquid Chromatography-Mass Spectrometry Methods in Food and Environmental Analysis, Imperial College Press, London.

[31] Kole, P. L., Venkatesh, G., Kotecha, J. & Sheshala, R. (2011). Recent advances in sample preparation techniques for effective bioanalytical methods. *Biomed. Chromatogr.*, 25, 199-217.

[32] Chen, Y., Guo, Z., Wang, X. & Qiu, C. (2008). Sample preparation. *J. Chromatogr. A*, 1184, 191-219.

[33] Ibrahim, N., Osman, R., Abdullah, A. & Saim, N. (2014). Determination of phthalate plasticisers in palm oil using online solid phase extraction-liquid chromatography (SPE-LC). *J. Chem.*, 682975/1-682975/10.

[34] Hao, L., Wang, C., Ma, X., Wu, Q., Wang, C. & Wang, Z. (2014). Magnetic three-dimensional graphene solid-phase extraction coupled with high performance liquid chromatography for the determination of phthalate esters in fruit juice. *Anal. Methods*, 6, 5659-5665.

[35] Gao, X., Yang, B., Tang, Z., Luo, X., Wang, F., Xu, H. & Cai, X. (2014). Determination of Phthalates Released from Paper Packaging Materials by Solid-Phase Extraction-High-Performance Liquid Chromatography. *J. Chromatogr. Sci.*, 52, 383-389.

[36] Luo, X., He, C., Zhang, F., Wang, H., Yang, B. & Liang, X. (2014). Heat-shrink tubing as a solid-phase microextraction coating for the enrichment and determination of phthalic acid esters. *J. Sep. Sci.*, 37, 3656-3661.

[37] Holadova, K., Prokupkova, G., Hajslova, J. & Poustka, J. (2007). Headspace solid-phase microextraction of phthalic acid esters from vegetable oil employing solvent based matrix modification. *Anal. Chim. Acta*, 582, 24-33.

[38] Chen, H., Liu, X. J., Yang, C., Gao, J., Ye, C. W. & Li, X. J. (2009). Determination of Phthalates in Beverages by Headspace SPME-GC Using Calix[6]arene Fiber. *Chromatographia*, 70, 883-890.

[39] Sun, M., Tang, R., Wu, Q., Wang, C. & Wang, Z. (2013). Graphene reinforced hollow fiber liquid-phase microextraction for the

determination of phthalates in water, juice and milk samples by HPLC. *Anal. Methods*, 5, 5694-5700.

[40] Farajzadeh, M. A., Aghdam, S. R., Afshar Mogaddam, M. R. & Ghorbanpour, H. (2015). An elevated temperature-dispersive liquid-liquid microextraction method combined with GC-flame ionization detection as a sensitive method for determining phthalate esters. *Anal. Methods*, Ahead-

[41] Viñas, P., Campillo, N., Pastor-Belda, M., Oller, A. & Hernández-Cordoba, M. (2015). Determination of phthalate esters in cleaning and personal care products by dispersive liquid-liquid microextraction and liquid chromatography-tandem mass spectrometry. *J. Chromatogr. A*, 1376, 18-25.

[42] Qi, P., Wang, J., Li, Y., Su, F., Jin, J. & Chen, J. (2011). Molecularly imprinted solid-phase extraction coupled with HPLC for the selective determination of monobutyl phthalate in bottled water. *J. Sep. Sci.*, 34, 2712-2718.

[43] Ma, Y., Hashi, Y., Ji, F. & Lin, J. M. (2010). Determination of phthalates in fruit jellies by dispersive SPE coupled with HPLC-MS. *J. Sep. Sci.*, 33, 251-257.

[44] Xie, Q., Liu, S., Fan, Y., Sun, J. & Zhang, X. (2014). Determination of phthalate esters in edible oils by use of QuEChERS coupled with ionic-liquid-based dispersive liquid-liquid microextraction before high-performance liquid chromatography. *Anal. Bioanal. Chem.*, 406, 4563-4569.

[45] Gómez-Hens, A. & Aguilar-Caballos, M. P. (2003). Social and economic interest in the control of phthalic acid esters. *TrAC, Trends Anal. Chem.*, 22, 847-857.

[46] Llompart, M., Garcia-Jares, C. & Landin, P. (2006). Phthalate esters. *Chromatogr. Sci. Ser.*, 93, 1103-1153.

[47] Barrera, P. B., Alonso, M. C. B., Feas, C. P., Vazquez, E. P. & Hermelo, P. H. (2010). Analytical methods for phthalates determination in biological and environmental samples: a review. *Bisphenol A Phthalates: Uses, Health Eff. Environ. Risks*, 29-61.

[48] Xu, D. P., Li, S., Chen, Y. H., Li, H. B., Li, A. N. & Xu, X. R. (2013). Phthalates: toxicity, occurrence and analytical methods. *Int. J. Mod. Biol. Med.*, 4, 12-29, 18.

[49] Method 606-Phthalate ester. Appendix A to part 136. Methods for organic chemical analysis of municipal and industrial wastewater. US Code Fed. Regul., Title 70 (1999).

[50] David, F., Sandra, P., Tienpont, B., Vanwalleghem, F. & Ikonomou, M. (2003), Analytical methods review, in: C.A. Staples (Ed.). Phthalate Esters, Collection: Handbook of Environmental Chemistry Series, vol 3, pp. 9-56.

[51] Gallart-Ayala, H., Núñez, O. & Lucci, P. (2013). Recent advances in LC-MS analysis of food-packaging contaminants. *TrAC, Trends Anal. Chem.*, 42, 99-124.

[52] Fankhauser-Noti, A. & Grob, K. (2007). Blank problems in trace analysis of diethylhexyl and dibutyl phthalate: Investigation of the sources, tips and tricks. *Anal. Chim. Acta*, 582, 353-360.

[53] Reid, A. M., Brougham, C. A., Fogarty, A. M. & Roche, J. J. (2007). An investigation into possible sources of phthalate contamination in the environmental analytical laboratory. *Int. J. Environ. Anal. Chem.*, 87, 125-133.

[54] P. Lucci, C.P.B. Martins. Liquid Chromatography-High Resolution Mass Spectrometry in Environmental and Food Analysis, Chapter 8, in: O. Núñez, H. Gallart-Ayala, C.P.B. Martins, P. Lucci (Eds), Fast Liquid Chromatography-Mass Spectrometry Methods in Food and Environmental Analysis, Imperial College Press, London, p. 325-345, 2015.

[55] Chen, D., Miao, H., Zou, J., Miao, H., Lu, L., Zhao, Y. & Wu, Y. (2015). Determination of Phthalate Esters in Liquor by High Resolution Mass Spectrometry. *Anal. Lett.*, 48, 739-751.

[56] Jia, W., Chu, X., Ling, Y., Huang, J. & Chang, J. (2014). Analysis of phthalates in milk and milk products by liquid chromatography coupled to quadrupole Orbitrap high-resolution mass spectrometry. *J. Chromatogr. A*, 1362, 110-118.

[57] Kaufmann, A. (2012). The current role of high-resolution mass spectrometry in food analysis. *Anal. Bioanal. Chem.*, 403, 1233-1249.

[58] Lucci, P. & Martins C.P.B.. Liquid Chromatography-High Resolution Mass Spectrometry in Environmental and Food Analysis, Chapter 8 (2015), in: O. Núñez, H. Gallart-Ayala, C.P.B. Martins, P. Lucci (Eds), Fast Liquid Chromatography-Mass Spectrometry Methods in Food and Environmental Analysis, Imperial College Press, London, p. 325-345.

[59] Al Saleh, I., Shinwari, N. & Alsabbaheen, A. (2011). Phthalates residues in plastic bottled waters. *J. Toxicol. Sci.*, 36, 469-478.

[60] Celeiro, M., Lamas, J. P., Garcia-Jares, C. & Llompart, M. (2015). Pressurized liquid extraction-gas chromatography-mass spectrometry

analysis of fragrance allergens, musks, phthalates and preservatives in baby wipes. *J. Chromatogr. A*, 1384, 9-21.

[61] Nair, J. J., Ndhlala, A. R., Chukwujekwu, J. C. & Van Staden, J. (2012). Isolation of di(2-ethylhexyl) phthalate from a commercial South African cognate herbal mixture. *S. Afr. J. Bot.*, 80, 21-24.

[62] Fierens, T., Servaes, K., Van Holderbeke, M., Geerts, L., De Henauw, S., Sioen, I. & Vanermen, G. (2012). Analysis of phthalates in food products and packaging materials sold on the Belgian market. *Food Chem. Toxicol.*, 50, 2575-2583.

[63] Myridakis, A., Balaska, E., Gkaitatzi, C., Kouvarakis, A. & Stephanou, E. G. (2015). Determination and separation of bisphenol A, phthalate metabolites and structural isomers of parabens in human urine with conventional high-pressure liquid chromatography combined with electrospray ionisation tandem mass spectrometry. *Anal. Bioanal. Chem.*, 407, 2509-2518.

[64] Li, X., Xiong, W., Lin, H., Zhuo, L., Lv, S., Tang, X., Chen, M., Zou, Z., Lin, Z., Qiu, B. & Chen, G. (2013). Analysis of 16 phthalic acid esters in food simulants from plastic food contact materials by LC-ESI-MS/MS. *J. Sep. Sci.*, 36, 477-484.

[65] Tranfo, G., Papaleo, B., Caporossi, L., Capanna, S., De Rosa, M., Pigini, D., Corsetti, F. & Paci, E. (2013). Urinary metabolite concentrations of phthalate metabolites in Central Italy healthy volunteers determined by a validated HPLC/MS/MS analytical method. *Int. J. Hyg. Environ. Health*, 216, 481-485.

[66] Kong, L., Kadokami, K., Wang, S., Duong, H. T. & Chau, H. T. C. (2015). Monitoring of 1300 organic micro-pollutants in surface waters from Tianjin, North China. *Chemosphere*, 122, 125-130.

[67] Lucci, P., Pacetti, D., Núñez, O. & Frega, N.G. (2012) Current Trends in Sample Treatment Techniques for Environmental and Food Analysis, Chapter 5. in: de Azevedo Calderon L. (Ed.), Chromatography - The Most Versatile Method of Chemical Analysis, InTech, p. 127-164.

[68] Li, Z., Huang, D., Fu, C., Wei, B., Yu, W., Deng, C. & Zhang, X. (2011). Preparation of magnetic core mesoporous shell microspheres with C18-modified interior pore-walls for fast extraction and analysis of phthalates in water samples. *J. Chromatogr. A*, 1218, 6232-6239.

[69] Wang, X., Song, G. & Deng, C. (2015). Development of magnetic graphene@hydrophilic polydopamine for the enrichment and analysis of phthalates in environmental water samples. *Talanta*, 132, 753-759.

[70] Pérez-Feás, C., Barciela-Alonso, M. C., Sedes-Díaz, A. & Bermejo-Barrera, P. (2010). Phthalates determination in pharmaceutical formulae used in parenteral nutrition by LC-ES-MS: importance in public health. *Anal. Bioanal. Chem.*, 397, 529-535.

[71] Pérez-Feás, C., Barciela-Alonso, M. C. & Bermejo-Barrera, P. (2011). Direct LC-ES-MS/MS determination of phthalates in physiological saline solutions. *J. Chromatogr. B: Anal. Technol. Biomed. Life Sci.*, 879, 231-235.

[72] Lu, L., Hashi, Y., Wang, Z. H., Ma, Y. & Lin, J. M. (2011). Determination of phthalate esters in physiological saline solution by monolithic silica spin column extraction method. *J. Pharm. Anal.*, 1, 92-99.

[73] Pérez-Feás, C., Barciela-Alonso, M. C. & Bermejo-Barrera, P. (2011). Presence of phthalates in contact lens and cleaning solutions. *Microchem. J.*, 99, 108-113.

[74] Li, W. & Duan, J. (2011). Detection of phthalates in water using ultra performance liquid chromatography-electrospray ionization tandem mass spectrometry MRM mode- "ghost peaks" and measurement methodology. *Anal. Methods*, 3, 314-321.

[75] Komesli, O. T., Bakirdere, S., Bayoeren, C. & Goekcay, C. F. (2012). Simultaneous determination of selected endocrine disrupter compounds in wastewater samples in ultra trace levels using HPLC-ES-MS/MS. *Environ. Monit. Assess.*, 184, 5215-5224.

[76] Khedr, A. (2013). Optimized extraction method for LC-MS determination of bisphenol A, melamine and di(2-ethylhexyl) phthalate in selected soft drinks, syringes, and milk powder. *J. Chromatogr. B: Anal. Technol. Biomed. Life Sci.*, 930, 98-103.

[77] Chang, L. W., Hou, M. L. & Tsai, T. H. (2013). Pharmacokinetics of dibutyl phthalate (DBP) in the rat determined by UPLC-MS/MS. *Int. J. Mol. Sci.*, 14, 836-849.

[78] Chen, B. & Zhang, L. (2013). An easy and sensitive analytical method of determination of phthalate esters in children's toys by UPLC-MS/MS. *Polym. Test.*, 32, 681-685.

[79] Chang-Liao, W. L., Hou, M. L., Chang, L. W., Lee, C. J., Tsai, Y. M., Lin, L. C. & Tsai, T. H. (2013). Determination and pharmacokinetics of di-(2-ethylhexyl) phthalate in rats by ultra performance liquid chromatography with tandem mass spectrometry. *Molecules*, 18, 11452-11466.

[80] Xu, D., Deng, X., Fang, E., Zheng, X., Zhou, Y., Lin, L., Chen, L., Wu, M. & Huang, Z. (2014). Determination of 23 phthalic acid esters in food by liquid chromatography tandem mass spectrometry. *J. Chromatogr. A*, 1324, 49-56.

[81] Hayasaka, Y. (2014). Analysis of phthalates in wine using liquid chromatography tandem mass spectrometry combined with a hold-back column: Chromatographic strategy to avoid the influence of pre-existing phthalate contamination in a liquid chromatography system. *J. Chromatogr. A*, 1372, 120-127.

[82] Commission Decision 2002/657/EC of 12 of August 2002 implementing Council Directive 96/23/EC concerning the performance of analytical methods and the interpretation of results, Off. J. Eur. Commun. L 221 (2002) 8.

[83] Petrovic, M. & Barceló, D. (2000). Determination of Anionic and Nonionic Surfactants, Their Degradation Products, and Endocrine-Disrupting Compounds in Sewage Sludge by Liquid Chromatography/Mass Spectrometry. *Anal. Chem.*, 72, 4560-4567.

[84] Hauser, R. & Calafat, A. M. (2005). Phthalates and human health. *Occup. Environ. Med.*, 62, 806-818.

[85] Hernández-Díaz, S., Mitchell, A. A., Kelley, K. E., Calafat, A. M. & Hauser, R. (2009). Medications as a potential source of exposure to phthalates in the U.S. population. *Environ. Health Perspect.*, 117, 185-189.

[86] Albro, P. W., Thomas, R. & Fishbein, L. (1973). Metabolism of diethylhexyl phthalate by rats. Isolation and characterization of the urinary metabolites. *J. Chromatogr.*, 76, 321-330.

[87] Albro, P. W. & Moore, B. (1974). Identification of the metabolites of simple phthalate diesters in rat urine. *J. Chromatogr.*, 94, 209-218.

[88] Wittassek, M., Koch, H. M., Angerer, J. & Bruening, T. (2011). Assessing exposure to phthalates - The human biomonitoring approach. *Mol. Nutr. Food Res.*, 55, 7-31.

[89] Kato, K., Silva, M. J., Needham, L. L. & Calafat, A. M. (2005). Determination of 16 Phthalate Metabolites in Urine Using Automated Sample Preparation and On-line Preconcentration/High-Performance Liquid Chromatography/Tandem Mass Spectrometry. *Anal. Chem.*, 77, 2985-2991.

[90] Frederiksen, H., Joergensen, N. & Andersson, A. M. (2010). Correlations between phthalate metabolites in urine, serum, and seminal plasma from young Danish men determined by isotope dilution liquid

chromatography tandem mass spectrometry. *J. Anal. Toxicol.*, 34, 400-410.

[91] Solymos, E., Guddat, S., Geyer, H., Flenker, U., Thomas, A., Segura, J., Ventura, R., Platen, P., Schulte-Mattler, M., Thevis, M. & Schaenzer, W. (2011). Rapid determination of urinary di(2-ethylhexyl) phthalate metabolites based on liquid chromatography/tandem mass spectrometry as a marker for blood transfusion in sports drug testing. *Anal. Bioanal. Chem.*, 401, 517-528.

[92] Jeong, J. Y., Lee, J. H., Kim, E. Y., Kim, P. G. & Kho, Y. L. (2011). Determination of phthalate metabolites in human serum and urine as biomarkers for phthalate exposure using column-switching LC-MS/MS. *SH W*, 2, 57-64.

[93] Frederiksen, H., Aksglaede, L., Sorensen, K., Skakkebaek, N. E., Juul, A. & Andersson, A. M. (2011). Urinary excretion of phthalate metabolites in 129 healthy Danish children and adolescents: Estimation of daily phthalate intake. *Environ. Res.*, 111, 656-663.

[94] Gries, W., Ellrich, D., Kuepper, K., Ladermann, B. & Leng, G. (2012). Analytical method for the sensitive determination of major di-(2-propylheptyl)-phthalate metabolites in human urine. *J. Chromatogr. B: Anal. Technol. Biomed. Life Sci.*, 908, 128-136.

[95] Dewalque, L., Pirard, C., Dubois, N. & Charlier, C. (2014). Simultaneous determination of some phthalate metabolites, parabens and benzophenone-3 in urine by ultra high pressure liquid chromatography tandem mass spectrometry. *J. Chromatogr. B: Anal. Technol. Biomed. Life Sci.*, 949-950, 37-47.

[96] Wang, H. x., Wang, B., Zhou, Y. & Jiang, Q. w. (2013). Rapid and sensitive analysis of phthalate metabolites, bisphenol A, and endogenous steroid hormones in human urine by mixed-mode solid-phase extraction, dansylation, and ultra-performance liquid chromatography coupled with triple quadrupole mass spectrometry. *Anal. Bioanal. Chem.*, 405, 4313-4319.

[97] Chen, M., Tao, L., Collins, E. M., Austin, C. & Lu, C. (2012). Simultaneous determination of multiple phthalate metabolites and bisphenol-A in human urine by liquid chromatography-tandem mass spectrometry. *J. Chromatogr. B: Anal. Technol. Biomed. Life Sci.*, 904, 73-80.

[98] Bao, J., Zeng, X. W., Qin, X. D., Lee, Y. L., Chen, X., Jin, Y. H., Tang, N. J. & Dong, G. H. (2015). Phthalate Metabolites in Urine Samples

from School Children in Taipei, Taiwan. *Arch. Environ. Contam. Toxicol.*, Ahead-

[99] Chang, Y. J., Lin, K. L. & Chang, Y. Z. (2013). Determination of Di-(2-ethylhexyl)phthalate (DEHP) metabolites in human hair using liquid chromatography-tandem mass spectrometry. *Clin. Chim. Acta*, 420, 155-159.

[100] Sanchis, Y., Coscolla, C., Roca, M. & Yusa, V. (2015). Target analysis of primary aromatic amines combined with a comprehensive screening of migrating substances in kitchen utensils by liquid chromatography-high resolution mass spectrometry. *Talanta*, 138, 290-297.

[101] Dirwono, W., Nam, Y. S., Park, H. M. & Lee, K. B. (2013). LC-TOF/MS determination of phthalates in edible salts from food markets in Korea. *Food Addit. Contam., Part B*, 6, 203-208.

[102] Self, R. L. & Wu, W. H. (2012). Rapid qualitative analysis of phthalates added to food and nutraceutical products by direct analysis in real time/orbitrap mass spectrometry. *Food Control*, 25, 13-16.

In: Phthalates
Editor: Miguel Owens
ISBN: 978-1-63483-908-2


*Chapter 4*

# PROPOSAL OF NEW APPROACHES FOR EXPOSURE ASSESSMENT OF PHTHALATES: EXPOSURE ASSESSMENT OF PHTHALATES IN POLYMER MATERIALS

***Yukihiro Ouchi,* Fumitaka Maruyama and Shigehiko Fujimaki***
Consumer Testing Service Division, SGS Japan Inc.
Hodogaya-ku, Yokohama, Japan

## ABSTRACT

Exposure of phthalates in polymer materials was assessed by employing migration modeling with advanced numerical algorithms.

A modeling approach originally developed in Europe for assessing food contact materials for regulatory compliance was employed to assess the exposure risk of phthalates in public health.

Migration estimation of phthalates was compared with the empirical data; the results of the migration modeling were within the allowable tolerance, thus proving its applicability to any contact medium, including human skin. In this respect, this migration estimation scheme can also be used for human exposure modeling to assess the exposure risk of phthalates.

---

* To whom correspondence should be addressed. E-mail: yukihiro.ouchi@sgs.com.

The new technique described above would enable effective assessment of exposure risks.

# 1. INTRODUCTION

Most people are exposed to a wide range of different products or articles containing substances with safety concerns on a day-to-day basis. Substances with safety concerns can be present in some final products and articles at a level that can cause harm to human health. These toxic substances can be transferred to human body by migration or emission through food products and ambient air or directly through dermal contact with the products. The amount of substances transferred varies depending on the conditions of use of the final products or articles, such as time, temperature, and contact frequency. The probable exposure pathways are not only food products but also indoor air and dermal contact with different polymer materials. The amount of exposure to these substances from all possible pathways forms the basis for exposure assessment.

Relevant exposure varies with season, geography, and lifestyle or ethnic background. The amount of substances transferred from polymeric materials and articles contacting can vary widely. The theory of diffusion can be used to mathematically estimate the quantity of substances transferred.

Recently, adverse effect of phthalic acid di-esters (phthalates) on human health has become a key issue of growing concern. Phthalates are used widely in industry as plasticizers and can be present in high concentrations in polymer-based products. However, some phthalates are now classified as major hazardous substances, posing a threat to the reproductive system (birth defects or other problems related to reproductive), and are suspected to have endocrine-disrupting properties [1, 2]. From this perspective, restrictions on these compounds are being regulated in some areas such as medical- and child-care products, and even stricter controls are expected to be enforced [3-8].

Several test methods for phthalates have been studied and standardized over the past couple of years [9-15]. However, these test methods are aimed at determining the content level of phthalates and cannot be used to directly estimate the amount of exposure. Content percentage needs to be converted to exposure amount by considering all possible combinations of exposure scenarios. From the viewpoint of consumer protection, the focus has to be on the health risks due to substances possibly released from the final products and articles. To this end, a realistic evaluation of the health risk associated with phthalates in polymeric materials and articles would be required.

Under Proposition 65, the Safe Drinking Water and Toxic Enforcement Act of 1986 (California), di(2-ethylhexyl)phthalate (DEHP) and four other phthalates, namely butyl benzyl phthalate (BBP), di-isodecyl phthalate (DIDP), di-n-butyl phthalate (DBP), and_di-n-hexyl phthalate (DnHP), are defined to have developmental toxicity or reproductive toxicity by Cal/EPA [16, 17]. For example, DEHP is listed as a chemical causing both cancer and birth defects with "no significant risk level (NSRL)" at 310 (μg/day) and "maximum allowable dose level (MADL)" of 4200 (μg/day) for adults, and 600 (μg/day) for infant boys.

The goal of this study is to propose a new method to assess the amount of exposure to phthalates. This can be achieved by examining the application of a modeling approach that has been validated and backed up by experiments conducted in Europe on food contact materials (FCMs). The purpose of this study is to demonstrate practical use of migration modeling for estimating exposure to phthalates. The aforementioned modeling approach enables us to estimate any type of migration, for example, long-term exposure or for some specific contact medium such as human skin, which would not be possible via experimental investigation.

## 2. BASIC THEORY—MIGRATION MODELING

Because a diffusion flow J can be defined as $J = -D \cdot \text{grad}(C)$, where D is the specific diffusion coefficient and C is the concentration of a migrant, the diffusion of a migrant in a divergence field can be expressed as

$$\partial C/\partial t = -\text{div}(J) = \text{div}(D \cdot \text{grad}(C) \quad (1)$$

Here, equation (1), more commonly known in the form of the following expressions, corresponds to the well-known formula of Fick's second law of diffusion.

$$\partial C/\partial t = D\nabla^2 C \quad (2)$$

$$\partial C/\partial t = D\,\partial^2 C/\partial x^2 \quad (3)$$

Migration of a substance in a polymer can be estimated on the basis of Fick's theory with a specific diffusion coefficient D. In addition, substance outflow from the boundary between the polymer and the contact medium is presumed to depend on the partition coefficient.

In other words, simulation of the migration process discussed in this article is based on Fick's second law of diffusion considering the partition between adjacent layers or contact media in closed systems [18].

Diffusion coefficients of substances in polymers are estimated using generally recognized estimation procedures such as the "Piringer model" based on polymer-specific constants (AP-values), which have been proven applicable to a few types of polymers in Europe [19]. In most cases of practical relevance, the diffusion process within the contact material is the rate-determining step of the migration process.

Estimation of partition coefficients between the contact layer and the contact medium is based on the migrants' polarity expressed by the octanol/water partition coefficient of the substances [20, 21]. The partitioning process (relationship between solubility of a substance in the material and contacting medium) strongly influences the overall amount of migration of a substance.

Time-dependent migration curves and concentration profiles have been computed for the migrant in both the material and the contact medium.

By employing advanced finite element analysis migration modeling, one can predict [22] the migrating amounts of organic substance such as residual monomers, additives, contaminants, reaction products, unintentionally added substances from plastic multi-layer, and multi-material multi-layer materials into packed food or other contact media such as pharmaceuticals, cosmetics, drinking water, or the environment.

For example, the verification of food packaging compliance by application of recognized diffusion models was legislated in Europe recently. The presented method ensures compliance of plastic FCMs and articles with specific migration limits according to EU Regulations and Swiss Legislation.

# 3. Results and Discussion

## Experimental Confirmation of Migration Modeling

The modeling method described here has been designed to simulate the overall diffusion process of a migrant in a polymer material. It does not necessarily reflect any individual behavior at the molecular level. Therefore, confirmation of its performance and limitation would be considered essential to ensure practical application to exposure assessment. With this in mind, experimental confirmation of migration modeling was performed as a preliminary step. A relatively simple geometrical setting was selected for the verification experiment from the viewpoint of convenience.

A commercially available cylindrical PVC tube containing 10 wt% DEHP was prepared and filled with 95% ethanol as the contact medium for 4 h under a constant temperature of 25°C. Both ends of the 8-cm-long PVC hollow tube were covered with a phthalate-free film. Figure 1 shows a schematic overview of the verification experiment.

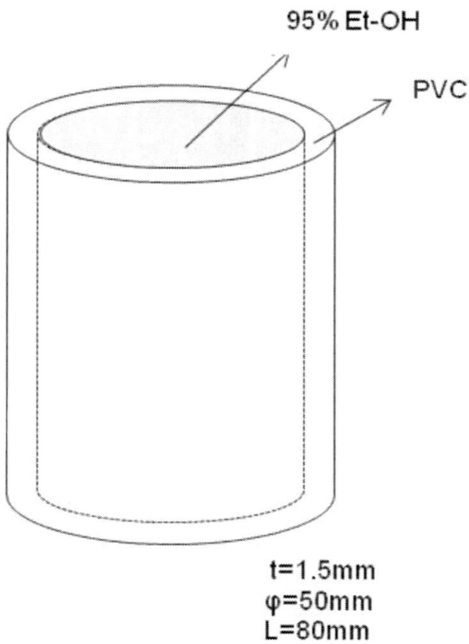


Figure 1. Schematic overview for the verification experiment.

A small amount of the contact medium was taken by a microsyringe once every hour for 4 h to actually measure the change in concentration of the extracted phthalate with Gas chromatograph-mass spectrometer (GC-MS). Moreover, migration modeling was performed in parallel by reflecting actual conditions in the calculation parameters to the extent possible, for example, geometry and layer designs, polymer selection, substance and stimulant, diffusion and partition coefficient setting, and other conditions such as temperature and processing time. Then the observed concentration values were plotted on the time-dependent simulation curve to assess the effectiveness of migration modeling. For this particular case, the Brandsch equation [23] was selected as the most suitable diffusion coefficient capable of plausible estimation in harmony with actual measured values.

Figure 2 shows typical computer-generated migration behavior of phthalates in comparison with experimental results.

A reasonably good match for both the time-dependent migration curves and the concentration profile was found with the experimental results. These profiles, in comparison with experimental values, indicate that the effectiveness of migration modeling is acceptable, especially in cases where experimental investigation is difficult or unavailable.

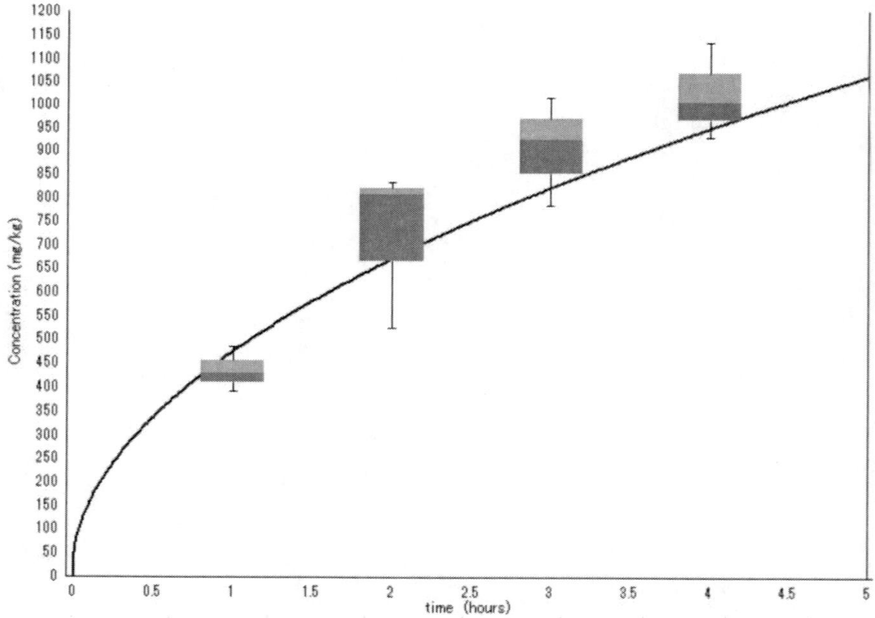


Figure 2. Typical computer-generated migration behavior of phthalates in comparison with experimental results (each block and bar on the graph represent the minimum, maximum, mean and statistical areas of 4 repeated measurements).

## Application of Migration Modeling to Exposure Assessment of Phthalates

As given above, migration modeling was proven valid empirically within an allowable tolerance, and its application to human exposure assessment of phthalates in polymer materials was examined by applying the Brandsch equation as the most suitable diffusion coefficient for migration modeling of phthalate exposure process.

Although the modeling method was initially developed and released to ensure compliance of plastic FCMs and articles with specific migration limits according to EU Regulations and Swiss Legislation, it has the potential to simulate migration in any contact medium.

By inputting relevant parameters, the migration model can estimate exposure of a substance in any contact medium, including food, air, solids, saliva, perspiration, human skin, and mucous membranes.

For each migrant, the simulation will require input of the partition coefficient between the last layer and the contact medium and the diffusion coefficient in the contact medium.

Most probably, partition coefficient of the migrant in human skin would be complex and undetermined. Considering that skin is an organic medium, a starting partition coefficient of K = 1 would be reasonable (as the worst-case scenario). However, human skin can be a good barrier against migration; therefore, a partition coefficient of K = 1000 would be reasonable as well. Real partition coefficients between PVC and human skin should be in the range of K = 10–100, for example, K = 50.

Diffusion coefficients in the human skin are expected to be in the range of 1e-8 to 1e-10 $cm^2/s$ at room temperature. A diffusion coefficient of 1e-9 $cm^2/s$ was selected as the representative value for typical exposure to human skin.

Figure 3 shows the condition setting for migration modeling based on the assumption of human skin contact.

A cylindrical human skin stimulant, equivalent to human body density of 1.07 g/cm3, was selected as a substitute for the human arm. The geometrical setting, therefore, is based on the assumption that a 1-mm-thick PVC sheet is wrapped around the human arm. Migration modeling was performed with an initial DEHP concentration of 0.5% at 27°C (skin temperature) for 4 h.

Figure 4 shows the exposure risks of dermal contact based on migration modeling of DEHP under a partition coefficient of K = 50.

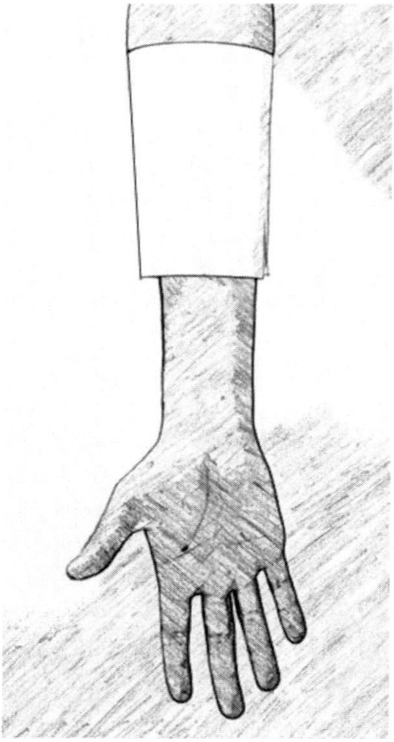

Figure 3. Condition setting of migration modeling based on the assumption of human skin contact.

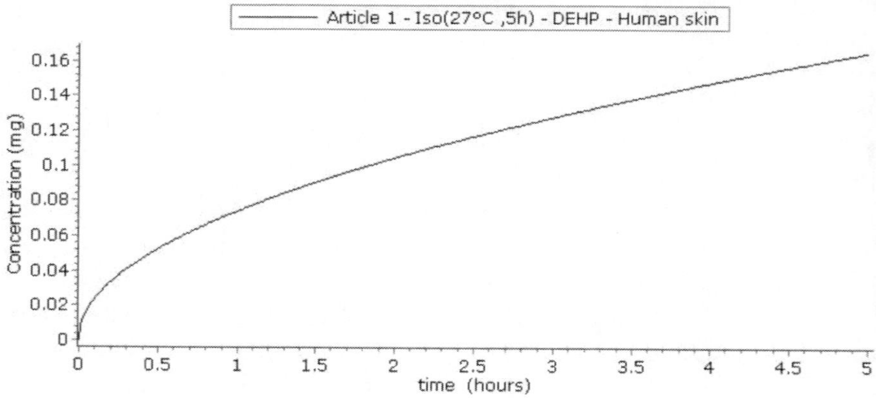


Figure 4. Exposure risk of dermal contact based on migration modeling of phthalate (DEHP) under a partition coefficient of K = 50.

Assuming that real partition coefficients can be in the range of K = 1–1000, most likely in the range of K= 10–100, the simulation curve shown in Figure 4 was almost as expected with presumable concentration values. These results indicate the applicability of this modeling approach to any contact medium, including human skin. In this respect, this migration estimation method can most likely be used for modeling human exposure to phthalates.

## CONCLUSION

Migration behavior of phthalates in polymer materials was estimated by employing an advanced theoretical diffusive model.

Experimental confirmation of migration modeling delivered a good example within an allowable tolerance and indicated its applicability to any contact medium, including human skin.

In this regard, this modeling approach holds promise as a method for assessing the level of human exposure to phthalates. Based on this positive outlook, further research would lead to the establishment of an effective procedure for assessing the risk of exposure to phthalates.

## ACKNOWLEDGMENTS

The authors are grateful to Mr. Y. Miyata and K. Yokoyama of the Consumer Testing Service Division, SGS Japan Inc. for their thoughtful suggestions and helpful discussions.

## REFERENCES

[1] J. L. Lyche, A. C. Gutleb, A. Bergman, A. J. Murk, E. Ropstad, M. Saunders and J. U. Skaare (2009). Reproductive and Developmental Toxicity of Phthalates. *Journal of Toxicology and Environmental Heath*, Part B, Vol.12, Issue 4, 225 – 249.

[2] G. Bylina (2011). DANGEROUS PHTHALATES IN CHILDREN'S ENVIRONMENT. *Ecological Chemistry and Engineering*. S, Vol.18, No.4, 455 - 463.

[3] *Official Journal of the European Union*, "DIRECTIVE 2005/84/EC OF THE EUROPEAN PARLIAMENT AND OF THE COUNCIL of 14 December 2005 - restrictions on the marketing and use of certain dangerous substances and preparations (phthalates in toys and childcare articles).

[4] *Official Journal of the European Union*, "REGULATION (EC) No 1907/2006 OF THE EUROPEAN PARLIAMENT AND OF THE COUNCIL of 18 December 2006 concerning the Registration, Evaluation, Authorization and Restriction of Chemicals (REACH)," 2006.

[5] Section 108 of the Consumer Product Safety Improvement Act of 2008 (CPSIA).

[6] Official Journal of the European Union, "Directive 2011/65/EU of the European Pariliament and of the Council," 2011.

[7] *Official Journal of the European Union*, "Commission Delegated Directive (EU) 2015/863 of 31 March 2015 amending Annex II to Directive 2011/65/EU of the European Parliament and of the Council as regards the list of restricted substances," 2015.

[8] EN 14372:2004 Child use and care articles. Cutlery and feeding utensils -- Safety requirements and tests.

[9] T. Yuzawa, C. Watanabe, R. R. Freeman, and S. Tsuge (2009). Rapid and Simple Determination of Phthalates in Plastic Toys by a Thermal Desporption-GC/MS Method. Analytical Sciences, Vol.*25*, 1057 - 1058.

[10] Test Method: CPSC-CH-C1001-09.3 Standard Operating Procedure for Determination of Phthalates, April 1st, 2010.

[11] L. Huang, Z. Liu, L. Yi, C. Liu and D. Yang (2011). Determination of the Banned Phthalates in PVC Plastic Toys by the Soxhlet Extraction-Gas Chromatography/Mass Spectrometry Method. *International Journal of Chemistry*, Vol.*3*, No.2, 169 – 173.

[12] ASTM D7823 - 14.
Standard Test Method for Determination of Low Level, Regulated Phthalates in Poly (Vinyl Chloride) Plastics by Thermal Desorption—Gas Chromatography /Mass Spectrometry, 2013.

[13] ISO 8124-6:2014 Safety of toys – Part 6: Certain phthalates esters in toys and children's products.

[14] ISO14389:2014 Textiles – Determination of the phthalate content – Tetrahydrofuran method.

[15] F. Maruyama, S. Fujimaki, Y. Sakamoto, Y. Kudo and H. Miyagawa (2015). Screening of Phthalates in Polymer Materials by Pyrolysis GC/MS. *Analytical Sciences*, Vol.31, 3 – 5.

[16] Reproductive and Cancer Hazard Assessment Section Office of Environmental Health Hazard Assessment California Environmental Protection Agency, "Chemicals under Consideration for Possible Listing via the Authoritative Bodies Mechanism," PACKAGE 21, 2005.

[17] Reproductive and Cancer Hazard Assessment Section Office of Environmental Health Hazard Assessment California Environmental Protection Agency, "Chemicals under Consideration for Possible Listing via the Authoritative Bodies Mechanism," PACKAGE 6d, 2003.

[18] C. Simoneau, ed., "Application of generally recognized diffusion models for the estimation of specific migration in support of EU Directive 2002/72/EC," *JRC Scientific and Technical Reports*, 2010.

[19] Official Journal of the European Union, "European Commission Regulation (EU) No 10/2011 relating to plastic materials and articles intended to come into contact with foodstuffs," 2011.

[20] D. Jenke (2011). A General Assessment of the Pysiochemical Factors That Influence Leachables Accumulation in Pharmaceutical Drug Products and *Related Solutions. PDA* Journal of Pharmaceutical Science and Technology, Vol.54, 166 – 176.

[21] A. Ozaki, A. Gruner, A. Störmer, R. Brandsch and R.Franz (2010). Correlation between Partition Coefficients Polymer/Food Simulant, KP,F, and Octanol/Water, Log POW – a New Approach in support of Migration Modeling and Compliance Testing. DLR (Deutsche Lebensmittel-Rundschau), 106 Jahrgang, 203 – 208.

[22] B. Roduit, C. H. Borgeat, S. Cavin, C. Fragniere and V. Dudler (2005). Application of Finite Element Analysis (FEA) for the simulation of release of additives from multilayer polymeric packaging structures. *Food Additives and Contaminations*, Vol.22, No.10, 945 – 955.

[23] Rainer Brandsch, to be published.

# INDEX

## D

## E

## F



## G



## H

## M



## N

## Q



## R



## S

## T



## U

## V



## W



## X